Marx Joyce
Abbott Hardy Emerson Austen
Defoe Melville Machiavelli Cooper Hugo
Montaigne Chesterton Eliot Grimm
Haggard
Stoker Carroll Molière
Wilde Christie Maupassant Byron Schiller
Garnett Fitzgerald Engels
Goethe Einstein Hawthorne Smith Kafka
Cotton Dostoyevsky Hall
Baum Kipling Doyle Willis
Henry Nietzsche
Leslie Dumas Flaubert Turgenev Balzac
Stockton Vatsyayana Crane
Burroughs Verne
Curtis Tocqueville Vinci
Homer Widger Tolstoy Gogol Busch
Darwin Whitman
Thoreau Twain Scott
Potter Freud Zola Plato Harte
Kant Jowett Stevenson Lawrence Dickens Hesse
Andersen Burton
London Descartes Cervantes
Poe Aristotle Wells Voltaire
Hale James Hastings Cooke
Bunner Shakespeare Irving
Richter Chambers
Doré Dante da Benedict Alcott
Swift Chekhov Shaw Wodehouse
Pushkin
Newton

tredition

tredition was established in 2006 by Sandra Latusseck and Soenke Schulz. Based in Hamburg, Germany, tredition offers publishing solutions to authors and publishing houses, combined with worldwide distribution of printed and digital book content. tredition is uniquely positioned to enable authors and publishing houses to create books on their own terms and without conventional manufacturing risks.

For more information please visit: www.tredition.com

TREDITION CLASSICS

This book is part of the TREDITION CLASSICS series. The creators of this series are united by passion for literature and driven by the intention of making all public domain books available in printed format again - worldwide. Most TREDITION CLASSICS titles have been out of print and off the bookstore shelves for decades. At tredition we believe that a great book never goes out of style and that its value is eternal. Several mostly non-profit literature projects provide content to tredition. To support their good work, tredition donates a portion of the proceeds from each sold copy. As a reader of a TREDITION CLASSICS book, you support our mission to save many of the amazing works of world literature from oblivion. See all available books at www.tredition.com.

 Project Gutenberg

The content for this book has been graciously provided by Project Gutenberg. Project Gutenberg is a non-profit organization founded by Michael Hart in 1971 at the University of Illinois. The mission of Project Gutenberg is simple: To encourage the creation and distribution of eBooks. Project Gutenberg is the first and largest collection of public domain eBooks.

The Myxomycetes of the Miami Valley, Ohio

A. P. (Andrew Price) Morgan

Imprint

This book is part of TREDITION CLASSICS

Author: A. P. (Andrew Price) Morgan
Cover design: Buchgut, Berlin – Germany

Publisher: tredition GmbH, Hamburg - Germany
ISBN: 978-3-8472-1624-7

www.tredition.com
www.tredition.de

Copyright:
The content of this book is sourced from the public domain.

The intention of the TREDITION CLASSICS series is to make world literature in the public domain available in printed format. Literary enthusiasts and organizations, such as Project Gutenberg, worldwide have scanned and digitally edited the original texts. tredition has subsequently formatted and redesigned the content into a modern reading layout. Therefore, we cannot guarantee the exact reproduction of the original format of a particular historic edition. Please also note that no modifications have been made to the spelling, therefore it may differ from the orthography used today.

From the Journal of the Cincinnati Society of Natural History,
Oct. 1892, Jan. 1893.

THE MYXOMYCETES OF THE MIAMI VALLEY, OHIO.

By A. P. Morgan.

First Paper.

(Read January 3, 1893.)

Preston, Hamilton County, Ohio, December 28, 1892.

Mr. Davis L. James

Dear Sir—Along with this I send you the first installment of the papers, entitled "The Myxomycetes of the Miami Valley, Ohio."

The work in these papers is based upon my ample collection of Myxomycetes growing in this region, comprising more than one hundred species; these have been diligently compared with specimens obtained from correspondents elsewhere in this country and in Europe.

At the same time, I have also included many extra limital species. This has been done chiefly to more clearly elucidate the subject in places where the local material is not sufficient.

The only apology I can make for the arrangement which I present, is that I have been obliged to choose from several different systems. I have aimed not to hamper myself, by attaching paramount importance to some particular character throughout.

I purpose to furnish a synopsis of the whole at the end of the work.

Very truly yours,
A. P. Morgan.

MYXOMYCETES, Wallr.

Fructification essentially a minute membranaceous vesicle, the SPORANGIUM inclosing the SPORES, the product of a motile protoplasmic body called the PLASMODIUM.

Microscopic organisms with the habit of the Fungi. The ripe spore of the Myxomycetes is globose or ellipsoidal in shape, with the epispore colorless or colored, and smooth or marked by characteristic surface—sculpture according to the species; the spore in germination gives rise to an elongated protoplasmic body, which exhibits amoeboid movements, and is known by the name of *swarm-cell*. The swarm-cells [Pg 2] multiply by bipartition, which may be repeated through several generations; they then unite together to form the large motile protoplasmic bodies named *plasmodia*. The newly-formed plasmodium is distinguished by its greater size from the swarm-cells, while it exhibits essentially the same movements and changes of shape. The plasmodia gradually increase in size, and as they grow assume commonly the form of branched strands; these spread over the surface of the substratum, which is usually the decaying parts of plants, in the form of veins and net-works of veins, giving rise to a copiously-branched reticulated or frill-like expansion, which covers surfaces varying in extent from a few to several centimeters. They are chiefly composed of a soft protoplasm of the consistence of cream, which may be readily spread out into a shapeless smear, and is usually colorless, but sometimes exhibits brilliant colors of yellow, orange, rose, purple, etc. The development of the plasmodium ceases with the formation of the *spores* within their *sporangia*.

The formation of the sporangia out of the plasmodium appears under three general forms, which, however, pass into each other and are, therefore, not strictly limited.

First: An entire plasmodium spread out on its substratum becomes transformed into a sporangium, or it divides into a variable number of unequal and irregular pieces, each of which undergoes transformation. Such a sporangium lying flat on the substratum, more or less elongated and flexuous, often branched and reticulate, is termed a *plasmodiocarp*.

Second: Erect sporangia on a narrow or stalk-like base, begin as node-like swellings on the branches of the plasmodium, and gradually rise to their ultimate form as the surrounding protoplasm flows into them and assumes an upward direction. These sporangia are nearly always perfectly regular in shape; they may be globose, obovoid, somewhat depressed, or more or less elongated, and are either stipitate or sessile.

Third: A number of plasmodia collect together from every side and become fused into a single body, often of considerable dimensions; from these combinations originate the large spore-receptacles which are called *æthalia*. The component sporangia may be regular in shape, standing close together, in a single stratum, with entire connate walls; more [Pg 3] often, being elongated and flexuous, they branch and anastomose freely, their walls becoming perforated and more or less defective; in other cases, the æthalium is a compound plasmodiocarp, the narrow sinuous sporangia branched and anastomosing in all directions, forming an intricate network, closely packed together and inseparable. The surface of the æthalium is often covered by a continuous layer of some excreted substance, which is called the *common cortex*.

The wall of the sporangium, typically, is a thin, firm membrane, colorless and pellucid, or colored in various shades of violet, brown, yellow, etc.; it is sometimes extremely delicate, as in Lamproderma, or is scarcely evident, as in Stemonitis; in other instances it is thickened by deposits on the inner surface, as in Tubulina, or by incrustations on the outer surface, as in Chondrioderma. The stipes are tubes usually with a thick wall, which is often wrinkled and folded lengthwise, and is confluent above with the wall of the sporangium; in some cases the stipe also enters the sporangium, and is more or less prolonged within it as a *columella*. The stipe commonly expands at the base into a membrane, which fastens it to the substratum, and is called the *hypothallus*; when all the stipes of the same group of sporangia stand upon a single continuous membrane, it is called a *common hypothallus*.

In the simplest forms, the cavity of the sporangium is filled exclusively with the numerous spores; but in most all of the genera, tubules or threads of different forms occur among the spores and constitute the *capillitium*. The capillitium first makes its appearance in Reticularia, in which upon the inner surface of the walls of the sporangia there are abundant fibrous thickenings; next in Cribraria it is spread over the inner surface of the wall, and is early separated from it; here, also, it first assumes a more definite form and arrangement; in Physarum it is in connection with the wall of the sporangium only by its extremities while it traverses the interior with a complicated network; in Stemonitis and its allies the capillitium originates wholly from the columella; in most species of Arcyria it issues from the interior of the stipe. The capillitium in Trichia consists of numerous slender threads which are *free*, that is, are not attached in any way; they are usually simple and pointed at each extremity; the surface of these threads exhibits beautiful spiral markings. [Pg 4]

Order I. LICEACEÆ.

Sporangia always sessile, simple and regular or plasmodiocarp, sometimes united into an æthalium. The wall a thin, firm, persistent membrane, often granulose-thickened, usually rupturing irregularly. Spores globose, usually some shade of umber or olivaceous, rarely violaceous.

The species of this order are the simplest of the Myxomycetes; the sporangium, with a firm, persistent wall contains only the spores. There is no trace of a capillitium, unless a few occasional threads in the wall of Tubulina prefigure such a structure. To the genera of this order is appended the anomalous genus Lycogala, which seems to me better placed here than elsewhere.

Table of Genera of Liceaceæ.

- 1. Licea. Sporangia simple and regular or plasmodiocarp, gregarious; hypothallus none.
- 2. Tubulina. Sporangia cylindric, or by mutual pressure becoming prismatic, distinct or more or less connate and æthalioid, seated upon a common hypothallus.
- 3. Lycogala. Æthalium with a firm membranaceous wall; from the inner surface of the wall proceed numerous slender tubules, which are intermingled with the spores.

I. LICEA, Schrad. Sporangia sessile, simple and regular or plasmodiocarp, gregarious, close or scattered; hypothallus none; the wall a thin, firm membrane, sometimes thickened with scales or granules, breaking up irregularly and falling away or dehiscent in a regular manner. Spores globose, variously colored.

The sporangia are not seated on a common hypothallus; they are, consequently, more or less irregularly scattered about on the substratum. [Pg 5]

1. Licea variabilis, Schrad. Plasmodiocarp not much elongated, usually scattered, sometimes closer and confluent, somewhat depressed, the surface uneven or a little roughened and not shining, reddish-brown or blackish in color; the wall a thin, firm pellucid membrane, covered by a dense outer layer of thick brown or blackish scales, rupturing irregularly. Spores in mass pale ochraceous, globose or oval, even or nearly so, 13–16 mic. in diameter.

Growing on old wood. Plasmodiocarp 1–1.5 mm. in length, though sometimes confluent and longer. The wall is thick and rough, not at all shining. It is evidently the species of Schweinitz referred to by Fries under this name.

2. Licea Lindheimeri, Berk. Sporangia sessile, regular, globose, gregarious, scattered or sometimes crowded, dark bay in color, smooth and shining; the wall a thin membrane with a yellow-brown outer layer, opaque, rupturing irregularly. Spores in mass bright bay, globose, minutely warted, opaque, 5–6 mic. in diameter.

Growing on herbaceous stems sent from Texas. Sporangia about .4 mm. in diameter. The bright bay mass of spores within will serve to distinguish the species. The thin brown wall appears dark bay with the inclosed spores.

3. Licea biforis, Morgan, n. sp. Sporangia regular, compressed, sessile on a narrow base, gregarious; the wall thin, firm, smooth, yellow-brown in color and nearly opaque, with minute scattered granules on the inner surface, at maturity opening along the upper edge into two equal parts, which remain persistent by the base. Spores yellow-brown in mass, globose or oval, even, 9–12 mic. in diameter. See Plate III, Fig. 1.

Growing on the inside bark of Liriodendron. Sporangia .25-.40 mm. in length, shaped exactly like a bivalve shell and opening in a similar manner. I have also received specimens of this curious species from Prof. J. Dearness, London, Canada.

4. Licea Pusilla, Schrad. Sporangia regular, sessile, hemispheric, the base depressed, gregarious, chestnut-brown, shining; the wall thin, smooth, dark-colored and nearly opaque, dehiscent at the apex into regular segments. Spores in the mass blackish-brown, globose, even, 16-18 mic. in diameter. [Pg 6]

Growing on old wood, Sporangium about 1 mm. in diameter. On account of the color of the spores the genus *Protoderma* was created for this species by Rostafinski. It is number 2,316 of Schweinitz's N. A. Fungi.

II. TUBULINA. Pers. Sporangia cylindric, or by mutual pressure becoming prismatic, distinct or more or less connate and æthalioid, the apex convex, seated upon a common hypothallus; the wall a thin membrane, minutely granulose, firm and quite persistent, gradually breaking away from the apex downward. Spores abundant, globose, umber or olivaceous.

The sporangia usually stand erect in a single stratum, with their walls separate or grown together: in the more compact æthalioid forms, however, the sporangia, becoming elongated and flexuous, pass upward and outward in various directions, branching and anastomosing freely. See Plate III, Figs. 2, 3, 4.

1. Tubulina cylindrica, Bull. Sporangia cylindric, more or less elongated, closely crowded, distinct or connate, pale umber to rusty-brown in color, seated on a well developed hypothallus; the wall thin, firm, with minute veins and granules, semi-opaque, pale umber, often iridescent. Spores in mass pale umber to rusty-brown, globose, most of the surface reticulate, 6-8 mic. in diameter.

Growing on old wood, mosses, etc. Æthalium circular or irregular in shape, from one to several centimeters in extent, the individual sporangia 2-4 mm. in height. Plasmodium at first milky-white, soon changing to bright red, then to umber, becoming paler when mature and dry.

2. **Tubulina casparyi**, Rost. Sporangia more or less elongated, closely crowded and prismatic, connate, pale umber to brown in color, seated on a conspicuous hypothallus; the wall thin, firm, minutely granulose, semi-opaque, pale umber, iridescent when well matured; all or many of the sporangia traversed by a central columella, from which a few narrow bands of the membrane stretch to the adjacent walls. Spores in the mass pale umber to brown, globose, the surface reticulate, 7–9 mic. in diameter.

Growing on old prostrate trunks. Æthalium two or three to several centimeters in extent, the individual sporangia 3–5 [Pg 7] mm. in height. Plasmodium white, the immature sporangia dull-gray tinged with sienna color. The columella, with its radiating bits of membrane, is the same substance as the wall; it may be a reëntrant edge of the prismatic sporangium, caused by excessive crowding together; at least, this may be regarded as its origin; there may have arisen some further adaptation. The species is *Siphoptychium Casparyi*, Rost. I am indebted to Dr. George A. Rex for the specimens I have examined.

3. **Tubulina cæspitosa**, Peck. Sporangia short-cylindric, closely crowded, distinct or connate, argillaceous olive to olive-brown in color, seated on a well-developed hypothallus; the wall a thin membrane, with a dense layer of minute dark-colored round granules on the inner surface. Spores argillaceous olive in the mass, globose, minutely warted, 6–8 mic. in diameter.

Growing on old wood. Æthalium in irregular patches sometimes several centimeters in extent, the single sporangia about 1 mm. in height. Plasmodium dark olivaceous, the sporangia blackish if dried when immature, taking a paler shade of olivaceous, according to development and maturity. This is *Perichæna cæspitosa*, Peck, in the 31st N. Y. Report.

III. **LYCOGALA**. Mich. Æthalium with a firm membranaceous wall; from the inner surface of the wall proceed numerous slender tubules, which are intermingled with the spores. The material of the wall appears under three different forms: the inner layer is a thin membrane, uniform in structure, of a yellow-brown color, and semi-pellucid; the outer layer consists of large flat roundish or irregular vesicles, brown in color, filled with minute granules, and arranged in one or more strata; from these vesicles originate the tubules, which traverse the wall for a certain distance, and then enter the interior among the spores; the tubules are more or less compressed, simple or branched, and the surface is ornamented with warts and ridges, which sometimes form irregular rings and reticulations.

If the sporophores in this genus be regarded as simple sporangia, which is the view that Rostafinski takes of one of [Pg 8] the species, the tubules are simply the peculiar threads of a capillitium. If, however, the æthalium is a compound plasmodiocarp, the tubules stand for the original plasmodial strands and, consequently, represent the component sporangia.

1. Lycogala conicum. Pers. Æthalia small, ovoid-conic, gregarious, sometimes close together with the bases confluent, the surface pale umber or olivaceous marked with short brown lines, regularly dehiscent at the apex. The wall thin; the outer layer not continuous, the irregular brown vesicles disposed in angular patches and elongated bands, which have a somewhat reticulate arrangement. The tubules appear as a thin stratum upon the inner membrane; they do not branch, and they send long slender simple extremities inward among the spores. Spores in mass pale ochraceous, globose, minutely warted, 5-6 mic. in diameter. See Plate III, Fig. 5.

Growing on old wood. Æthalium 2-5 mm. in height, the tubules 3-8 mic. in thickness. This is *Dermodium conicum* of Rostafinski's monograph, but the structure is essentially the same as in the other species. Massee evidently did not have specimens of this species. I have never seen any branching of the tubules either in the wall or in the free extremities of the interior.

2. Lycogala exiguum, Morg. n. sp. Æthalia small, globose, gregarious, the surface dark brown or blackish, minutely scaly, irregularly dehiscent. The wall thin; the vesicles with a dark polygonal outline, disposed in thin irregular reticulate patches, which are more or less confluent. The tubules appear as an interwoven fibrous stratum upon the inner membrane; they send long slender branched extremities inward among the spores. Spores in mass pale ochraceous, globose, nearly smooth, 5-6 mic. in diameter. See Plate III, Fig. 6.

Growing on old wood. Æthalium 2-5 mm. in diameter, the threads 2-10 mic. in thickness, with very slight thickenings of the membrane. The polygonal vesicles give a reticulate appearance to the dark-brown patches which ornament the surface of the wall.

3. Lycogala epidendrum, Buxb. Æthalia subglobose, [Pg 9] gregarious, sometimes closely crowded and irregular, the surface umber, brown or olivaceous, minutely warted, at length, irregularly dehiscent at or about the apex. The wall thick, the brown vesicles loosely aggregated and densely agglutinated together, traversed in all directions by the much-branched tubules, which send long-branched extremities inward among the spores; the main branches thick and flat, with wide expansions, especially at the angles, the ultimate branchlets more slender and obtuse at the apex. Spores in the mass from pale to reddish ochre, globose, minutely warted, 5-6 mic. in diameter. See Plate III, Fig. 7.

Growing on old wood. Æthalium 5-12 mm. in diameter, the width of the tubules varying from 12-25 mic. in the main branches, with broader expansions at the angles, to 6-12 mic. in the more slender final branchlets. This is one of the most common of the Myxomycetes; it grows in all countries, and in this region may be found on old trunks at all seasons of the year.

4. Lycogala flavofuscum, Ehr. Æthalia large, subglobose or somewhat pulvinate, solitary or gregarious, the surface at first silvery-shining, becoming yellow-brown, minutely areolate, irregularly dehiscent. The wall very thick and firm, hard and rigid; the thick outer layer of roundish brown vesicles closely compacted in numerous strata; from the vesicles of the lower strata the long and broad much-branched tubules proceed into the interior among the spores; the ultimate branchlets clavate and obtuse at the apex. Spores in the mass pale ochre, cinerous or brownish, globose, minutely warted, 5-6 mic. in diameter. See Plate III, Figs. 8, 9.

Growing on old trunks. Æthalium 1 to several centimeters in diameter, the width of the tubules varying from 25-60 mic. in the main branches, with sometimes much broader expansions at the angles, to 10-25 mic. in the ultimate branchlets. The brown vesicles of the outer wall are easily separated from each other and emptied of their contents by maceration; it is then seen that a thin pellucid membrane incloses numerous roundish granules, much resembling the spores, but usually a little larger, 5-8 mic. in diameter. [Pg 10]

Order II.—RETICULARIACEÆ.

Sporangia simple, regular and stipitate, or compound, forming an æthalium; the wall a thin membrane with distinct fibrous thickenings upon the inner surface, the membrane, or at least certain portions of it, disappearing usually at the maturity of the spores, leaving behind the more permanent fibrous thickenings as a more or less definite capillitium. Spores globose, purple, brown, ochraceous, rarely violaceous.

In this order the threads of a capillitium first make their appearance; but they are confined to the inner surface of the wall of the sporangium, being set at liberty by the early decay of the outer membrane.

Table of Genera of Reticulariaceæ.

a. Æthalia.

- 1. Reticularia. Æthalium composed of numerous slender sinuous sporangia which repeatedly branch and anastomose.
- 2. Clathroptychium. Æthalium composed of numerous regular erect sporangia.

b. Sporangia simple.

- 3. Cribraria. Capillitium of slender threads combined into a network of polygonal meshes.
- 4. Dictydium. Capillitium of numerous convergent ribs, which extend from base to apex, and are united by fine transverse fibers, thus forming a network of rectangular meshes.

I. RETICULARIA, Bull. Æthalium composed of numerous slender sinuous sporangia, which repeatedly branch and anastomose, closely packed together and seated upon a com [Pg 11] mon hypothallus, the apices of the final branches coherent at the surface, and naked or covered by an additional corticate layer. Walls of the sporangia consisting of a thin membrane, with abundant fibrous thickenings, presenting broad expansions, narrowing to thin flat bands, and reduced in many places to slender fibrous threads. Spores abundant, globose, umber or violaceous.

After the maturity of the spores disintegration of the sporangial wall begins, the thin membrane disappearing more rapidly than the fibrous thickenings or the portions of the sporangial walls near the base, which are more compactly grown together; there is thus left at each stage an increasing number of the shreddy fibers mingled with the spores.

1. Reticularia Splendens, Morg. n. sp. Æthalium pulvinate, circular or more or less elongated and irregular, seated on a conspicuous silvery hypothallus; the surface naked, bright umber, smooth and shining. Walls of the sporangia firm and quite persistent, pale umber, slowly disintegrating, consisting for the most part of wide expansions, with their angles tapering to narrow bands and slender threads. Spores in the mass pale umber, globose, most of the surface reticulate, 7-9 mic. in diameter. See Plate III, Fig. 10.

Growing on old wood. Æthalium from 1 to several centimeters in extent and 5-10 mm. in thickness, usually growing singly, rarely close enough to be confluent. This species has lately been referred to *Reticularia rozeana*, Rost., but it varies greatly from the account given of that species in the Journal of Botany for September, 1891.

2. Reticularia Umbrina. Fr. Æthalium pulvinate, roundish, more or less irregular, the surface covered by a thin, silvery, shining, common cortex, which at the base is confluent with the hypothallus. Walls of the sporangia umber or rusty-brown next the base, with broad expansions in places thickly grown together, toward the surface passing into narrow bands and abundant fibrous threads, which rapidly disintegrate. Spores in the mass umber or rusty brown, globose, most of the surface reticulate, 7-9 mic. in diameter. [Pg 12]

Growing on old trunks. Æthalium one to several centimeters in extent, and 5-15 mm. in thickness. The walls of the sporangia are much more reduced to the shreddy fibrous condition than in the preceding species, and on this account they much more rapidly disintegrate, causing the æthalium soon to collapse. It is *Reticularia Lycoperdon*, Bull.

3. Reticularia atra, A. & S. Æthalium pulvinate, variable in form and size, covered with a thin, fragile, blackish, cortical layer. Walls of the sporangia violaceous, next the base with broad expansions, in places more thickly grown together, toward the surface becoming narrow with more abundant fibrous threads, sometimes presenting a loose irregular network, the whole structure, however, quite variable, according to the stage of the disintegration. Spores globose, violet, minutely warted, 14–16 mic. in diameter.

Growing on wood and bark, especially of pine. Æthalium 2 or 3 to several centimeters in extent. This is *Amaurochæte atra* of Rostafinski's monograph, but the structure appears to be altogether similar to that of *Reticularia umbrina*.

II. CLATHROPTYCHIUM, Rost. Æthalium composed of numerous regular erect sporangia, seated in a single compact stratum, on a well-developed hypothallus, the surface formed by the coherent apices. Sporangia at first cylindric, with the apex convex and the wall entire; soon, by mutual pressure, they become prismatic and the lateral faces disappear, leaving the edges and the apex permanent. Spores globose, ochraceous.

1. Clathroptychium rugulosum, Wallr. Æthalium composed of numerous very slender sporangia, closely compacted into a single stratum, and seated on a conspicuous silvery hypothallus; the surface ochroleucous, honey color or olivaceous. The sporangia are typically hexangular when the lateral faces disappear, leaving at the edges six simple triangular threads, extending from the angles of the hexagonal apex downward to the base. Spores in the mass ochraceous, yellowish or brownish, globose, minutely warted, 8–10 mic. in diameter. [Pg 13]

Growing on old wood. Æthalium somewhat circular, or often quite irregular in shape, 1 to several centimeters in extent, the individual sporangia nearly 1 mm. in height, but scarcely .1 mm. in thickness. Deviations from the typical form of the sporangia sometimes occur, they are not seldom pentangular, and I have seen the apices quadrangular, with only four threads, or even triangular, and with but three; the threads, too, are said occasionally to branch and anastomose. *Reticularia plumbea*, Fries, S. M. III, 88; and *Ostracoderma spadiceum*, Schw., N. A. Fungi No. 2,381.

III. CRIBRARIA, Pers. Sporangia simple, globose or obovoid, stipitate, often cernuous; the wall regularly thickened on the inner surface in two ways, the lower basal portion by radiating ribs consisting of minute brown granules, the upper part by slender threads combined into a network of polygonal meshes; the basal portion of the membrane is commonly persistent with its thickening and is called the *calyculus*, the upper part nearly always disappears from the network at maturity; there are usually nodules of the brown granules at the angles of the network. Spores globose, purple, brown, ochraceous.

a. Sporangium, large.

1. Cribraria argillacea, Pers. Sporangia globose or obovoid, stipitate or nearly sessile, standing close together on a thin and evanescent hypothallus; the wall quite firm, silvery-shining, the greater portion persistent, breaking away about the apex; calyculus small, the brown radiating ribs soon passing into a network of polygonal meshes, the threads with irregular granulose-thickened portions at intervals throughout their whole extent. Stipe very short, erect, brown. Spores in the mass argillaceous, globose, 5–7 mic. in diameter.

Growing in large irregular patches on rotten trunks. Sporangia .6-.8 mm. in diameter, the stipe always much shorter than the sporangium, sometimes nearly obsolete. The resemblance of this species to some forms of *Tubulina cæspitosa* is very great. [Pg 14]

2. Cribraria vulgaris, Schrad. Sporangium large, globose, stipitate, somewhat cernuous; the calyculus brown, finely ribbed and granulose within, occupying but a small part of the sporangium; the network of slender threads, with very small nodules at the angles, each with several (3–7) radiating threads, sometimes with one or two free extremities, the meshes triangular or rhombic. Stipe rather short, stout, tapering upward, usually a little bent or curved at the apex, dark purplish brown in color. Spores in the mass pale ochraceous, globose, even, 5–7 mic. in diameter.

Growing on old wood. Sporangium .5-.7 mm. in diameter, the stipe two or three times the diameter of the sporangium in length. Recognized by the large sporangium and the very small nodules with their few radiating threads.

3. Cribraria dictydioides, C. & B. Sporangium large, globose, stipitate, cernuous; the calyculus small, with thickish brown ribs, from which the outer thin membrane often disappears soon after maturity; the network of slender threads, with large brown nodules at the angles, more or less elongated and irregular in shape, each with numerous (5–15) radiating threads, usually some with free extremities, the meshes largely triangular. Stipe long, tapering upward, flexuous, curved at the apex, dark purplish-brown in color. Spores in mass pale ochraceous, globose, even, 5–7 mic. in diameter.

Growing on rotten wood, especially of oak. Sporangium .5–.6 mm. in diameter, the stipe from three to five times as long. This species appears to be intermediate between *Cribraria vulgaris* and *Cribraria intricata*; the nodules are usually large and irregular, but the characteristic parallel threads of *C. intricata* do not often occur. The outer membrane of the calyculus is by no means always absent.

4. Cribraria elegans, B. & C. Sporangium rather large, globose, stipitate, somewhat cernuous; the calyculus thickly coated inside with dark purple granules, faintly ribbed, occupying about a third part of the sporangium; the network of slender threads, with large irregular dark purple nodules, quite variable in shape and size, angular and lobed, below sometimes much elongated, the meshes very irregular. Stipe rather [Pg 15] short, tapering upward, bent at the apex, dark purple in color. Spores in the mass bright purple, globose, even, 5–7 mic. in diameter.

Growing on old wood. Sporangium .4–.5 mm. in diameter, the stipe two or three times as long. It does not appear to be greatly different from *Cribraria purpurea*, Schrad.

b. Sporangium, small.

5. Cribraria tenella. Schrad. Sporangium small, globose, stipitate, cernuous; the calyculus brown, shining, granulose within and faintly ribbed, occupying from one-fourth to one-half the sporangium, sometimes the outer thin membrane early disappearing; the network of slender threads with small roundish or irregular nodules at the angles, each with several (4–8) radiating threads, sometimes two or three with free extremities, the meshes triangular or rhombic. Stipe long, tapering upward, flexuous, curved at the apex, purplish-

brown in color. Spores pale ochraceous in mass, globose, even, 5–7 mic. in diameter.

Growing on old wood. The sporangium .3–.4 mm. in diameter, the stipe three to five times as long. This is a much more delicate species than *Cribraria dictydioides*. The calyculus is variable in size; in some examples the thin connecting membrane between the ribs has disappeared.

6. Cribraria microcarpa, Schrad. Sporangium very small, globose, stipitate, somewhat cernuous; the calyculus represented by a few short brown ribs, the outer membrane soon disappearing; the network of slender threads, with small roundish nodules at the angles, each with several (4–6) radiating threads, with an occasional free extremity, the meshes largely rhombic. Stipe very long, slender, somewhat flexuous, bent at the apex, purplish-brown in color. Spores in mass pale ochraceous, globose, even, 6–7 mic. in diameter.

Growing on old wood. Sporangium .22–.27 mm. in diameter, the stipes 1–2 mm. in length. Readily distinguished by its very small sporangium and the comparatively very long stem. I am indebted to Dr. George A. Rex for specimens of this species. [Pg 16]

7. Cribraria cuprea, Morg. n. sp. Sporangium very small, oval or somewhat obovoid, stipitate, cernuous; the calyculus copper-colored, finely ribbed and granulose within, occupying from one-third to one-half the sporangium; the network of slender threads, with rather large triangular or quadrilateral meshes, and with large irregular dark copper-colored nodules, each having several (4–7) radiating threads, with an occasional free extremity. Stipe not very long, tapering upward, curved at the apex, of the same color as the sporangium or darker below. Spores pale coppery in mass, globose, even, 6–7 mic. in diameter. See Plate III, Fig. 11.

Growing on old wood. Sporangium .30–.35 X .25–.30 mm, the stipe two to four times as long as the sporangium. A minute species, easily recognized by its almost uniform color of bright new copper.

IV. DICTYDIUM, Schrad. Sporangium simple, depressed-globose, stipitate, cernuous; the wall regularly thickened on the inner surface by numerous convergent ribs, which extend from base to apex and are united by fine transverse fibers, thus forming a network of rectangular meshes; the basal portion of the membrane sometimes persists as a calyculus, the upper part disappears at maturity. Spores globose; purplish.

The ribs run from base to apex like the meridians on a globe; they are simple, or here and there they separate into two divergent branches, which sometimes again converge into one; at the apex of the sporangium there is usually a small irregular net in which all the ribs terminate.

1. Dictydium cernuum, Pers. Sporangium depressed-globose, umbilicate at the apex, stipitate, cernuous, purplish-brown in color; the calyculus granulose within, occupying from one-fourth to one-third of the sporangium, the ribs united by firm, persistent fibers. Stipe not very long, erect, tapering upward, bent at the apex, purplish-brown, the apex pale and pellucid, standing on a small hypothallus. Spores purplish-brown in mass, globose, even, 5–7 mic. in diameter.

Growing on old wood. Sporangium .4–.5 mm. in diameter, the stipe two or three times longer than the diameter of the [Pg 17] sporangium. This appears to be the species figured and described by Rostafinski and by Massee.

2. Dictydium longipes, Morg. n. sp. Sporangium large, depressed-globose, the apex umbilicate, stipitate, cernuous, dark purple in color; calyculus usually wholly wanting, the ribs united by weak fibers, which are easily torn asunder, allowing the ribs to curl up inwards. Stipe very long, flexuous, tapering upward, curved and twisted at the apex, dark purple in color, standing on a thin hypothallus. Spores in the mass dark purple, globose, even, 5–7 mic. in diameter. See Plate III, Fig. 12.

Growing on rotten wood, mosses, etc. Sporangium .5–.7 mm. in diameter, the stipe three to five times as long. This is a much larger species than the preceding; it has a uniform dark purple hue, the stipe is very long and much bent and twisted, the ribs of the sporangium are soon torn apart and rolled inward. [Pg 18]

EXPLANATION OF PLATE III

- Fig. 1.—Licea biforis, Morgan, n. sp.

- Figs. 2, 3, 4. — Diagrammatic representation of the structure of Tubulina
- Fig. 5. — Lycogala conicum, Pers., natural size
- Fig. 6. — Lycogala exiguum, Morgan, n. sp., natural size
- Fig. 7. — Lycogala epidendrum, Buxb., natural size
- Fig. 8. — Lycogala flavofuscum, Ehr., natural size
- Fig. 9. — Portion of tubule of Lycogala flavofuscum
- Fig. 10. — Reticularia splendens, Morgan, n. sp., natural size
- Fig. 11. — Cribraria cuprea, Morgan, n. sp.
- Fig. 12. — Dictydium longipes, Morgan, n. sp.

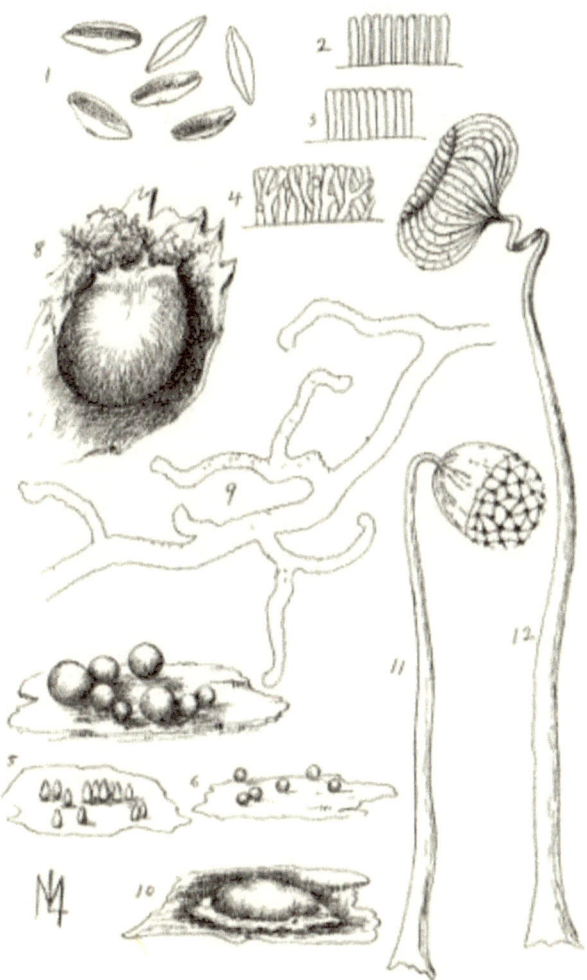

The Journal of the Cin. Soc. Natural History
Vol. XV. Plate III.

[Pg 19]

From the Journal of the Cincinnati Society of Natural History, April, 1893.

THE MYXOMYCETES OF THE MIAMI VALLEY, OHIO.

By A. P. Morgan.

Second Paper.

(Read May 2, 1893.)

Order III. PERICHÆNACEÆ.

Sporangia sessile or plasmodiocarp; the wall a thin membrane, with a more or less thickened outer layer of minute brownish scales and granules. Capillitium of long and very slender tubules, proceeding from numerous points of the sporangial wall, loosely branched, forming no evident network, the surface minutely warted or spinulose. Spores globose, oval, or somewhat irregular, yellow.

The order is distinguished by the sessile sporangia, with thick brown walls, and the very slender threads of the capillitium, with irregular and indefinite markings.

Table of Genera of Perichænaceæ.

- 1. Perichæna. Sporangia more or less depressed, roundish or more commonly polygonal and irregular, dehiscent in a circumscissile manner.
- 2. Ophiotheca. Plasmodiocarp terete and more or less elongated, bent and flexuous, sometimes annular or reticulate, irregularly dehiscent.

I. PERICHÆNA, Fr. Sporangia more or less depressed, roundish or more commonly polygonal and irregular, the edges approximate and sometimes confluent; the wall a thin [Pg 20] membrane, with a thick dense yellow-brown outer layer of minute scales and granules, becoming darker at the surface, dehiscent in a circumscissile manner. Capillitium of very slender loosely-branched threads, with the surface minutely warted. Spores globose, oval or somewhat irregular, yellow.

Distinguished from Ophiotheca by the flattened sporangium with a regular circumscissile dehiscence.

1. Perichæna depressa, Lib. Sporangia very much depressed, polygonal, irregular, crowded, the edges contiguous, sometimes confluent; the wall thick, yellow-brown within and scarcely impressed by the spores; the outer surface smooth, brown-red to brown or blackish in color, dehiscent in a circumscissile manner. Capillitium of slender loosely-branched threads, 1–3 mic. in thickness, the surface merely uneven or very minutely warted. Spores globose, yellow, 9–10 mic. in diameter. See Plate I, Fig. 13.

Growing on the inside of the bark of Juglans, Acer, etc. Sporangia variable in size, 7–1.3 mm. in breadth, irregular and angular, much flattened. It is said to include *Perichæna vaporaria*, Schw.

2. Perichæna irregularis, B. & C. Sporangia depressed, irregular, polygonal, crowded, the edges contiguous and sometimes confluent; the wall thick, yellow inside and faintly reticulately impressed by the spores, the outer surface smooth, purplish-brown, dehiscent in a circumscissile manner. Capillitium of slender-loosely branched threads, about 2 mic. in thickness, the surface minutely warted or spinulose. Spores subglobose, yellow, 9–10 mic. in diameter.

Growing on the outer bark of Acer, etc. Sporangium .5-.6 mm. in width, closely crowded and irregular. It is much smaller than *Perichæna depressa*, and its threads are more distinctly warted and spinulose.

3. Perichæna corticalis, Batsch. Sporangia globose, the base depressed, gregarious: the wall thick, yellow within and distinctly reticulately impressed by the spores, the outer surface reddish-brown or yellow-brown in color, dehiscent in a circumscissile manner. Capillitium of slender loosely-branched threads, about 2 mic. in thickness, the surface very minutely warted. Spores subglobose, yellow, 10–12 mic. in diameter. [Pg 21]

Growing on the inside of the bark of Elm. Sporangia .5-.6 mm. in diameter, quite regular in shape, with a slightly flattened base. My specimens are from Prof. McBride, of Iowa.

4. Perichæna marginata, Schw. Sporangia depressed, polygonal, approximate and sometimes confluent, the surface cinereous-

pulverulent, seated on a silvery hypothallus; the wall firm, thick, the outer surface yellow-brown, covered with minute whitish scales, the inner surface yellow, deeply reticulately impressed by the spores which rest against it, dehiscent in a circumscissile manner. Capillitium consisting of a few simple or somewhat branched threads or well-nigh obsolete. Spores subglobose, yellow, 12–14 mic. in diameter.

Growing on the outer surface of the bark of Acer, Fagus, etc. Sporangia .4-.6 mm. in width. This is plainly *Perichæna cano-flavescens*, Raunkier. I do not find any threads of a capillitium in my specimens.

II. OPHIOTHECA, Currey. Plasmodiocarp terete and more or less elongated, bent and flexuous, sometimes annular or reticulate, the surface not polished or shining: the wall a thin membrane, with a thin outer layer of minute scales and granules, irregularly dehiscent. Capillitium of very slender loosely-branched threads, with the surface minutely warted and spinulose. Spores globose, oval or somewhat irregular, yellow.

Distinguished from Perichæna by the terete plasmodiocarp and by the more spinulose capillitium. *Cornuvia* of Rostafinski.

1. Ophiotheca chrysosperma, Currey. Plasmodiocarp globose or oblong to elongated, and bent or flexuous, sometimes annular or branched and reticulate, dull brown in color; the wall a thin yellowish membrane, with a thin yellow-brown outer layer, irregularly dehiscent. Capillitium of slender loosely-branched threads, 2–3 mic. in thickness, the surface minutely spinulose. Spores subglobose, yellow, 8–9 mic. in diameter.

Growing on the inner surface of old bark of Quercus, etc. Plasmodiocarp .4-.5 mm. in thickness, variable in length. *Cornuvia circumscissa* of Rostafinski's monograph. [Pg 22]

2. Ophiotheca Wrightii, B. & C. Plasmodiocarp more or less elongated, bent and flexuous, very commonly in small rings, from brownish-ochre to brown or blackish in color, not polished; the wall a thin yellow membrane, with a thin brown outer layer, irregularly dehiscent. Capillitium of slender loosely-branched threads, 2–3 mic. in thickness, furnished with numerous straight or bent long-pointed

spinules. Spores subglobose, yellow, minutely warted, 10-12 mic. in diameter. See Plate I, Fig. 14.

Growing on the inside of bark of Acer, Carya, etc. Plasmodiocarp about .5 mm. in thickness, variable in length, often in small rings 1-2 mm. in diameter. The prickly threads are quite characteristic; the spinules are 3-5 mic. in length. *Hemiarcyria melanopeziza*, Speg., is evidently the same thing.

3. Ophiotheca vermicularis, Schw. Plasmodiocarp terete and more or less elongated, bent and flexuous, sometimes annular or reticulate, the surface not polished, brownish in color; the wall a thin yellow membrane, covered on the outside by a more or less thickened brown layer of scales and granules, irregularly dehiscent. Capillitium of slender loosely branched threads, 2-3 mic. in thickness, the surface with minute warts and ridges. Spores subglobose, yellow, 10-12 mic. in diameter.

Growing on the inside of old bark. Plasmodiocarp about .4 mm. in thickness and various in length; in my specimens the sporangia are mostly small rings. The species looks exactly like *Ophiotheca Wrightii*, but the character of the threads is quite different.

4. Ophiotheca pallida, B. & C. Plasmodiocarp terete, oblong or elongated annular and flexuous, the surface dull, pale ochraceous; the wall a thin pellucid membrane, minutely granulate, with a thin pale ochraceous outer layer, irregularly dehiscent. Capillitium of slender loosely-branched threads, 2-3 mic. in thickness, the surface minutely warted or spinulose. Spores subglobose, pale yellow, 10-12 mic. in diameter.

Growing on dead stems of herbaceous plants. Plasmodiocarp .3-.4 mm. in thickness, variable in length, sometimes short and roundish or oblong, sometimes much elongated and flexuous. More delicate than *Ophiotheca vermicularis*, and distinguished by its pallid color throughout. [Pg 23]

Order IV. ARCYRIACEÆ.

Sporangia regular and stipitate, rarely sessile; the wall a thin membrane, minutely granulose, colored as the spores and capillitium, the upper part soon torn away in a somewhat circumscissile manner, and early disappearing. Capillitium of slender tubules, repeatedly branching and anastomosing to form a complicated network of evident meshes, more or less expanded after dehiscence; the surface of the threads minutely warted or spinulose or with elevated ridges in the shape of rings, half rings or reticulations.

This order is specially distinguished by the threads of the capillitium forming a complicated network of evident meshes.

Table of Genera of Arcyriaceæ.

- 1. Lachnobolus. Capillitium of slender tubules, quite variable in thickness, proceeding from numerous points of the sporangial wall.
- 2. Arcyria. Capillitium of slender tubules, issuing from the interior of the stipe, the network without any free extremities.
- 3. Heterotrichia. Capillitium issuing from the interior of the stipe, the peripheral portion of the network bearing numerous short acute free branches.

I. LACHNOBOLUS, Fr. Sporangia stipitate or sessile, the wall a thin delicate membrane, minutely granulose, rupturing irregularly. Stipe short or sometimes wanting. Capillitium of slender tubules quite variable in thickness, proceeding from numerous points of the sporangial wall and forming a complicated network, the surface minutely warted or spinulose. Spores globose, yellowish or flesh-color. [Pg 24]

This genus differs from Arcyria in the capillitium springing from numerous points of the sporangial wall.

1. Lachnobolus globosus, Schw. Sporangia globose, stipitate, pale yellow, changing to clay-color; the wall thin and delicate, pellucid, minutely granulose, the upper part torn away and soon disappearing, the lower half more persistent. Stipe short, tapering upward, expanding at the base into a small hypothallus. Capillitium arising from the lower portion of the sporangium, forming a complicated network, the threads 3–5 mic. in thickness, the surface closely cov-

ered with minute warts. Spores globose, pale yellow to clay-color in mass, 8–9 mic. in diameter. See Plate I, Fig. 15.

Growing on the spines of Chestnut burs. Sporangia .5-.6 mm. in diameter, the stipe shorter than the sporangium.

2. Lachnobolus incarnatus, A. & S. Sporangia globose or ellipsoidal, substipitate, closely crowded and seated on a common hypothallus; the wall thin and delicate, pellucid, minutely granulose, dehiscing irregularly. Stipe very short or often obsolete. Capillitium proceeding from the inner surface of the sporangial wall, forming a complicated network, the threads extremely variable in thickness, minutely warted and spinulose. Spores globose, flesh-color in the mass, 8–9 mic. in diameter.

Growing on old wood. Sporangia .5-.8 mm. in height, sessile on a narrow base or with a very short stipe; the threads of the capillitium are generally 3–5 mic. in thickness, but there are broader expansions at the nodes and elsewhere. My specimens are from Prof. McBride, of Iowa. The species is extremely variable, and these specimens differ much from those described elsewhere.

II. ARCYRIA, Hill. Sporangia regular ovoid to cylindric, stipitate; the wall a thin delicate membrane, circumscissile or torn away near the base, the upper portion evanescent, the lower part persistent, small and cup-shaped. Stipe more or less elongated, the interior containing roundish vesicles which become smaller upward, and gradually pass into the normal spores. Capillitium of slender tubules, issuing from the interior of the stipe, forming a complicated network, without [Pg 25] any free extremities, the surface minutely warted or spinulose or with annular ridges. Spores globose, red, brown, yellow, cinereous.

§1. Clathroides, Mich. Capillitium closely attached by a few threads which issue from the interior of the stipe, and are free from the calyculus (except in *A. punicea*), much elongated after dehiscence, weak and drooping or prostrate; the meshes open and irregular, not differing externally and internally, their threads similar throughout, the warts or ridges of the surface exhibiting a spiral arrangement.

1. **Arcyria punicea,** Pers. Sporangium ovoid, more or less elongated; the calyculus small, plicate-sulcate. Stipe long, erect, brownish-red in color, expanded at the base into a small hypothallus. Capillitium firmly attached by numerous threads which are connate with the wall of the calyculus, much elongated after dehiscence, ovoid-oblong to cylindric, bright red in color, fading to red-brown or brownish-ochre; the threads uniform in thickness, about 3 mic., the surface with a series of prominent half-rings, which wind around the thread in a long spiral. Spores globose, even, 6–8 mic. in diameter.

Growing on old bark, wood, mosses, etc. The stipe 1–2 mm. in length, the capillitium elongated 2–4 mm. The commonest of the species, conspicuous by reason of its bright red color.

2. **Arcyria minor,** Schw. Sporangium ovoid-oblong; the calyculus small, sulcate and ribbed, granulose. Stipe short, erect, brownish-red in color, standing on a thin hypothallus. Capillitium much elongated after dehiscence, oblong to cylindric, lax and prostrate, bright red to brownish in color; the threads uniform in thickness, 2.5–3 mic., the surface with a series of prominent half-rings, which wind around the thread in a long spiral. Spores globose, even, 7–9 mic. in diameter. See Plate I, Fig. 17.

Growing on old wood, bark, Polyporus, etc. The stipe .4–.7 mm. in length, the capillitium elongated 1.5–3 mm. Not uncommon, but it is usually referred to *A. adnata.*

3. **Arcyria adnata,** Batsch. Sporangium ovoid; the calyculus very small, finely ribbed and granulose. Stipe very [Pg 26] short or entirely wanting. Capillitium much expanded after dehiscence, globose or obovoid, pale red to brownish in color; the threads uniform in thickness, about 4 mic., the surface with a series of prominent half-rings with mingled warts and spines, which wind around the thread in a long spiral. Spores globose, even, 6–8 mic. in diameter.

Growing in small clusters on old wood. A small species, the capillitium expanded 1–2 mm., the stipe extremely short, or altogether absent.

4. **Arcyria nutans,** Bull. Sporangium cylindric; the calyculus small, granulose, ribbed and sulcate. Stipe very short, arising from a

common hypothallus. Capillitium greatly elongated after dehiscence, cylindric, drooping and pendulous, pale yellow or pale ochraceous; the threads 3-4 mic. in thickness, the surface covered with spinules, among which are rings and half-rings, with an indistinct spiral arrangement. Spores globose, even, 7-9 mic. in diameter.

Growing on old wood. The capillitium elongated 4-8 mm., the stipe very short. A very conspicuous species by reason of its long pale yellow capillitium.

§2. Plectanella. Capillitium erect, firmly attached by numerous threads, which issue from the interior of the stipe, but are connate with the wall of the calyculus, after dehiscence not much expanded: the meshes at the surface of the network much smaller than those within, folded back and forth, narrow and irregular, their threads densely warted or spinulose; the meshes of the interior much larger, open and expanded, their threads with minute scattered warts or perfectly smooth.

5. Arcyria cinerea, Bull. Sporangium ovoid or oblong-ovoid; the calyculus very small. Stipe long, erect, cinereous, becoming blackish, standing on a thin hypothallus. Capillitium not much expanded after dehiscence, ovoid-oblong, erect, pale cinereous, sometimes pale yellowish; the external threads densely spinulose, 2-3 mic. in thickness; the threads of the interior thicker, 3-5 mic., and very minutely warted or quite smooth. Spores globose, even, 6-8 mic. in diameter.

Growing on old wood. Capillitium 1-2 mm. long, the stipe about the same length. [Pg 27]

6. Arcyria Cookei, Mass. Sporangium ovoid-cylindric, the calyculus very small. Stipe long, erect, gray to mouse-color, darker below, arising from a thin hypothallus. Capillitium not much expanded after dehiscence, ovoid-cylindric, erect, gray to mouse-color; the superficial threads densely and uniformly covered with minute warts, 3-5 mic. in thickness; the threads of the interior thinner, about 2 mic. and smooth, or with very minute scattered warts. Spores globose, even, 6-8 mic. in diameter. See Plate I, Fig. 16.

Growing on old wood, mosses, etc. Capillitium 1-2 mm. long, the stipe about the same length. It seems as common as *Arcyria cinerea*, and has heretofore been included in it. See Massee's Monograph, p. 154.

7. Arcyria digitata, Schw. Sporangium cylindric, the calyculus very small. Stipe long, ascending, brownish in color, usually several fasciculate or to some extent connate, the sporangia divergent at the apex. Capillitium not much expanded after dehiscence, cylindric, pale cinereous, or pale yellowish; the threads variable in thickness. 2–4 mic., those at the surface densely and minutely warted, those of the interior nearly smooth. Spores globose, even, 6–8 mic. in diameter.

Growing on old wood. Capillitium 2–4 mm. long, the stipe about the same length. *Arcyria bicolor*, B. & C.

III. HETEROTRICHIA, Massee. Sporangia regular, oblong-ovoid, stipitate; the wall a thin delicate membrane, the upper part disappearing at maturity, leaving the basal portion as a small calyculus. Stipe filled with large thick-walled vesicles, which are sub-angular from mutual pressure; these become smaller upward, and pass gradually into normal spores. Capillitium issuing from the interior of the stipe, the central and superficial threads dissimilar, forming a complicated network, with numerous free extremities, the surface minutely warted, or with annular ridges. Spores globose, brownish.

Distinguished from Arcyria by the numerous free extremities of the peripheral portion of the network. [Pg 28]

1. Heterotrichia Gabriellæ, Massee. Sporangium oblong-ovoid, stipitate; the calyculus small, thin, smooth. Stipe very short, erect, yellowish-brown in color. Capillitium much elongated after dehiscence, cylindric-ovoid, sub-erect; the threads of the central portion about 1.5 mic. thick, with slightly elevated ridges partly encircling the tube, nearly colorless; threads of the peripheral portion bright yellow, 5–6 mic. thick, with numerous short acute free branches, the surface densely and minutely warted. Spores in mass, yellowish-brown, globose, even, 7–8 mic. in diameter. See Plate I, Fig. 18.

Growing on wood; S. Carolina, *H. W. Ravenel*. The sporangia densely crowded, becoming scattered toward the margin of the cluster. Massee's Monograph of the Myxogasters.

Order V. TRICHIACEÆ.

Sporangium regular and stipitate or sessile, rarely plasmodiocarp; the wall a thin membrane, usually granular or venulose on the inner surface, colored as the spores and capillitium, irregularly dehiscent. Capillitium of slender tubules, simple or branched, scarcely forming an evident network; the surface of the threads furnished with continuous ridges, which wind around the tube in a spiral manner. Spores globose, red, brown, yellow, olivaceous.

This order is readily recognized by the spiral ridges which wind around the tubules of the capillitium.

Table of Genera of Trichiaceæ.

- 1. Hemiarcyria. Capillitium of long slender tubules, arising from the base of the sporangium, or issuing from the interior of the stipe; the spiral ridges parallel and conspicuous. [Pg 29]
- 2. Calonema. Capillitium of slender tubules, arising from the base of the sporangium; the surface traversed by a system of branching veins.
- 3. Trichia. Capillitium consisting of numerous short slender tubules, called elaters, which are wholly free; the spiral ridges parallel and conspicuous.
- 4. Oligonema. Capillitium scanty, composed of elaters habitually irregular and abnormal; the surface variously marked.

I. HEMIARCYRIA, Fr. Sporangia regular and stipitate, rarely plasmodiocarp, the wall at maturity breaking away from above downward, leaving more or less of the lower portion persistent. Stipe more or less elongated, rarely wanting, resting on a thin hypothallus. Capillitium of long slender tubules, more or less branched, arising from the base of the sporangium, or issuing from the interior of the stipe; the spiral ridges parallel and conspicuous, 3–5, rarely more in number, smooth or spinulose. Spores globose, red, yellow.

The genus is related on the one hand to Arcyria by the mode of attachment of the threads, on the other hand to Trichia, by the parallel spiral ridges which wind around them. By the mode of branching of the threads, the species fall readily into two sections.

§1. Arcyrioides. Capillitium of slender threads, branching and anastomosing, thus forming a more or less evident network.

In some of the species the large irregular meshes of the network are scarcely to be discerned, but are rather to be inferred from the abundant branching of the threads and the paucity of the free extremities.

1. Hemiarcyria plumosa, Morgan, n. sp. Sporangium obovoid to turbinate, olive-yellow to olive-brown in color, stipitate; the wall densely granulose within, externally smooth and shining, the upper part soon disappearing, leaving a funnel-shaped persistent base. Stipe long, erect, reddish-brown, arising from a thin hypothallus. Capillitium of threads 5-7 mic. in thickness, repeatedly branched and anastomosing, to form a dense network without any free extremities, [Pg 30] olive-yellow to olive-brown in color; the spiral ridges five or six, close, smooth. Spores in mass, lemon-yellow, globose, very minutely warted, 8-9 mic. in diameter. See Plate I, Fig. 19.

Growing gregariously on old damp logs; very common in this region. Sporangium with the stipe 2-3 mm. in height, the stipe usually much longer than the sporangium; the capillitium expands considerably after the disappearance of the upper part of the sporangium. This species is an Arcyria in every respect, except the spiral ridges, which wind about the thread of the capillitium.

2. Hemiarcyria Varneyi, Rex. Sporangium elongated ovoid, pale yellow, stipitate; the upper part of the wall disappearing at maturity, leaving a small cup-shaped persistent base. Stipe very short, dull brown. Capillitium of very slender threads 3.2-3.5 mic. in thickness, dull ochre in color, forming a network of small meshes, with numerous short slightly clavate free extremities, which proceed from the peripheral meshes; the spiral ridges seven or eight, winding unevenly, those of the superficial threads minutely spinulose. Spores in mass pale yellow, globose, even, 6-7 mic. in diameter.

Growing on old wood; Kansas, May Varney. Sporangium with the stipe about 1 mm. in height, the stipe very short. Dr. Rex, in

3. **Hemiarcyria ablata, Morgan n. sp.** Sporangium obovoid to turbinate, yellow or olive-yellow, stipitate; the wall rather firm, smooth and shining, breaking away about the apex, leaving the greater portion persistent. Stipe short, erect, yellow-brown to blackish in color, arising from a thin hypothallus. Capillitium of threads, 5-7 mic. in thickness, yellowish-ochre in color, more or less branched; the free extremities very scarce, obtuse or slightly swollen; the spiral ridges four or five, close, smooth or very minutely warted. Spores in mass, yellow, globose, minutely warted, 8-9 mic. in diameter.

Growing on old wood of Elm, etc. Sporangium with the stipe 1.5-2.5 mm. in height, the stipe variable in length, but not longer than the sporangium, diameter of the sporangium [Pg 31] .6-.8 mm. A half dozen threads proceed from the inner wall of the stipe branch twenty-five or thirty times, and afford scarcely half a dozen free ends.

4. **Hemiarcyria stipata, Schw.** Sporangia terete, elongated and flexuous, closely packed together and lying upon one another, stipitate, from bright incarnate to brick red or bay in color, smooth and shining; the wall thin and fragile, soon disappearing, except a small cup-shaped portion at the base. The stipes very short, often entirely concealed by the dense mass of sporangia, arising from a common hypothallus. Capillitium of threads somewhat variable in thickness, 3-6 mic., repeatedly branched and forming a network of very unequal meshes, with occasional clavate free extremities, pale to dark red in color; the spiral ridges three or four, often irregular, thickened or interrupted by minute warts and spinules. Spores in mass incarnate to brownish-red, globose, even, 7-9 mic. in diameter.

Growing on old wood of Liriodendron. Sporangia usually in small patches, each 1-2 mic. in length, the stipe very thin and short.

§2. **Hemitrichia.** Capillitium of very long slender threads, simple or remotely branched, and not forming a network, their further extremities all free.

The threads of the capillitium in these species are usually much coiled and entangled, but when straightened out they are seen to be very long, but few in number, fixed at one end and free at the other.

5. Hemiarcyria longifila, Rex. Sporangium obovoid or pyriform, yellow, stipitate; the wall a thin pellucid membrane, smooth and shining, beautifully iridescent, breaking away above the middle, the lower cup-shaped portion persistent. Stipe very short, reddish-brown to blackish, arising from a common hypothallus. Capillitium of slender threads, 3.5–4 mic. in thickness, golden yellow in color, simple or very rarely branched; the free extremities obtuse or slightly swollen, sometimes minutely apiculate; the spiral ridges, three or four, rather distant, with very minute scattered spinules or nearly smooth. Spores in mass, golden-yellow, globose, minutely warted, 9–10 mic. in diameter. [Pg 32]

Growing on old wood of Oak, etc. Sporangium with the stipe .8–1.5 mm. in height, the stipe very short, not exceeding the diameter of the sporangium. A small species, distinguished by its golden-yellow spores and capillitium.

6. Hemiarcyria funalis, Morgan n. sp. Sporangium obovoid to turbinate, yellow or olive yellow, polished stipitate; the wall firm, thickened on the inner surface by an olivaceous layer, breaking away from above downward, leaving an irregular cup-shaped base. Stipe short, reddish-brown to blackish, arising from a thin hypothallus. Capillitium of threads 6–8 mic. in thickness, yellowish-ochre or dull ochre in color, simple or remotely branched; the free extremities obtuse or swollen; the spiral ridges four or five, minutely warted. Spores in mass yellow, globose, minutely warted, 8–9 mic. in diameter. See Plate I, Fig. 20.

Growing on old wood. Sporangium 1.5–2.5 mm. in height, the stipe variable, but usually much shorter than the sporangium. Scarcely to be distinguished from *Hemiarcyria ablata*, except by the threads of the capillitium.

7. Hemiarcyria rubiformis, Pers. Sporangium obovoid or turbinate to cylindric, usually few to many fasciculate upon the united stipes, sometimes sessile, brown-red to brown or blackish in color, smooth and often shining with a metallic luster; the wall much thickened by a dense brownish-red layer of minute granules, at

maturity the apex torn away, leaving much the greater part persistent. Capillitium of slender threads, 4–6 mic. in thickness, brownish-red in color, very rarely branched; the free extremities usually terminated by a stout spine; the spiral ridges three or four, furnished with numerous spinules. Spores in mass, brownish-red, globose, minutely warted, 9–11 mic. in diameter.

Growing on old wood; one of the commonest of the Myxomycetes. The fascicle 3–4 mm. in height, the individual sporangia .5-.6 mm. in diameter.

8. **Hemiarcyria serpula**, Scop. Plasmodiocarp terete, flexuous, usually branching and anastomosing to form an extensive network, from tawny to golden-yellow in color; the wall thin above and yellow, breaking open irregularly and falling away down to the brownish thicker adherent base. [Pg 33] Capillitium consisting of a few long slender threads with numerous scattered short branches, the threads 4–6 mic. in thickness, golden-yellow; the free ends of the branches terminating in a slender spine; the spiral ridges three or four, covered with numerous slender spinules. Spores in the mass golden-yellow, globose, the surface reticulate, 10–12 mic. in diameter.

Growing on and inside of rotten wood. Plasmodiocarp an irregular patch, one to several centimeters in extent, the strands of the net about .5 mm. in thickness. A single reticulate plasmodium is usually converted without change of form into an individual plasmodiocarp.

II. CALONEMA, Morgan, gen. nov. Sporangia subglobose, irregular, sessile, without a hypothallus; the wall thin, marked with branching veins, irregularly dehiscent. Capillitium of slender tubules, arising from the base of the sporangium, repeatedly branched and with numerous free extremities; the surface traversed by a system of branching veins, ending in minute veinlets, which appear as irregular rings and spirals. Spores subglobose, yellow.

The habit of the single species is that of an Oligonema, and it has spores similar to those of most species of this genus, but the threads are long and branched, and they are fastened below to the base of the sporangium.

1. **Calonema aureum**, Morgan n. sp. Sporangia subglobose to turbinate, sessile, closely crowded and from mutual pressure quite irregular; the wall thin, marked with branching veins, golden-yellow in color, smooth and shining. Capillitium of threads more or less branched, 5-6 mic. in thickness, golden-yellow; the surface minutely venulose, and with larger rings and spirals, and sometimes with scattered spinules; the free extremities obtuse. Spores subglobose, yellow, the surface with elevated ridges combined into a network, 14-16 mic. in diameter. See Plate I, Fig. 21.

Growing on and within rotten wood. Sporangia quite irregular and variable in size, .3-.6 mm. in diameter. The beautiful venation of the wall of the sporangium is continued upon the surface of the threads of the capillitium. [Pg 34]

III. TRICHIA, Haller. Sporangia regular and stipitate or sessile and somewhat irregular; the wall, at maturity, irregularly ruptured. The stipe more or less elongated or often wanting, usually resting on a hypothallus. Capillitium consisting of numerous short slender tubules, called *elaters*, intermingled with the spores and wholly free; elaters simple or rarely branched a time or two, each extremity terminating in a smooth tapering point; the spiral ridges parallel and conspicuous, 2-5 in number, smooth or spinulose. Spores globose, yellow, ochraceous, olivaceous.

The genus Trichia is unique among the Myxomycetes in having its capillitium composed of tubules, which are entirely free from the wall of the sporangium. The length of these free tubes varies usually between .3 mm. and .5 mm., being sometimes shorter, but seldom longer; they are typically cylindric, or equally thickened from end to end, or quite rarely they are thickened in the middle, and taper gradually to each extremity; the extremities terminate in a smooth tapering point, straight or sometimes a little curved or flexuous, which maintains an average length in each species. The spiral ridges wind around the thread almost invariably to the left, or with the hands of a watch; they are always more or less prominent and conspicuous, and usually maintain a regular curve and uniform interval between each other in the same species; their surface is either smooth, or sometimes it is invested with minute warts or spinules.

In all the species of this genus, however, irregular and abnormal elaters are occasionally met with among the typical ones. As these abnormal forms always arrest attention, and have been conceived to possess specific value, it may be well to note the principal of them.

1. The elater is sometimes branched. In two or three species the branching appears to be quite regular and not abnormal; still, even in these species, most of the elaters in the sporangia are not branched. In some cases the branching arises from confluence of two or more elaters.

2. Ellipsoidal swellings, or enlargements of the elater, sometimes occur, at one or both extremities, or at points intermediate between them; these always occur irregularly, and are essentially abnormal. [Pg 35]

3. The smooth tapering point is rarely wanting, in which case the extremity presents a blunt end, the spiral ridges running to the end. More frequently the tapering points are multiplied, the elaters bearing two or three spines at the extremities; this often occurs in the species of Trichia, and also of Hemiarcyria with spinulose elaters.

4. The spiral ridges are sometimes defective, there being less than the typical number; sometimes they are merely displaced, there being a much wider interval between them than usual; rarely do they habitually wind about the thread in an irregular manner.

5. Under high magnifying power, fine ridges are sometimes seen running lengthwise of the elaters, bridging the intervals between the spirals. These were first observed by DeBary, in *Trichia chrysosperma*, but they have since been seen in the elaters of nearly every other species of Trichia, and also in species of Hemiarcyria.

The few species with elaters, so far as yet known, habitually irregular, defective and abnormal, are referred to the genus Oligonema.

The normal species of Trichia arrange themselves quite naturally into three sections.

§1. A nactium. Sporangia varying from globose to pyriform or turbinate, supported on a more or less elongated stipe. Spores globose, the surface minutely warted.

a. Elaters with very long tapering extremities.

1. Trichia fragilis, Sow. Sporangia obovoid to pyriform or clavate, often fasciculate, stipitate; the wall a thin membrane, with a thick dense outer layer of brown-red granules. Stipes long, erect or curved, simple or usually fasciculate and often connate, arising from a thin hypothallus. Mass of spores and capillitium from reddish-brown to yellow and ochraceous; elaters simple, rarely branched, 4-5 mic. thick, with very long tapering extremities, ending in smooth points 8-12 mic. long; spirals, three or four, perfectly smooth. Spores globose, minutely warted, 10-12 mic. in diameter. [Pg 36]

Growing on old wood. Sporangia with the stipe 2-4 mm. in height, the sporangium .6-.8 mm. in diameter, the stipe usually longer than the sporangium. The color quite variable, mostly dull red-brown or blackish-brown, more rarely yellow or coffee-brown, usually opaque, rarely shining.

2. Trichia fallax, Pers. Sporangium obovoid to pyriform or turbinate, rarely clavate, stipitate; the wall thin, smooth and shining, colored as the spores and capillitium. Stipe more or less elongated, simple, erect, brownish below, filled with roundish vesicles. Mass of capillitium and spores yellowish, ochraceous or olivaceous; elaters simple or sometimes with several branches, 4-6 mic. thick in the middle, tapering gradually to each extremity, ending in smooth tapering points, 20-40 mic. in length; spirals, three, perfectly smooth. Spores globose, minutely warted, 10-12 mic. in diameter. See Plate I, Fig. 22.

Growing on old wood. Sporangium with the stipe 2-4 mm. in height, sporangium .6-.8 mm. in diameter, the stipe usually longer than the sporangium. Under high magnifying power the spores are seen to be minutely reticulated.

b. Elaters cylindric, ending in a smooth tapering point.

3. Trichia subfusca, Rex. Sporangium globose, rarely globose-turbinate, stipitate; the wall thickish, dull tawny-brown above, shading to dark brown at the base. Stipe simple, erect, brown or blackish in color. Mass of capillitium and spores bright yellow; elaters simple, rarely branched, cylindric, 3.5-4 mic. in thickness, end-

ing in smooth tapering points, 10–12 mic. in length; spirals, four in number, perfectly smooth. Spores globose, minutely warted, 11.5–12.5 mic. in diameter.

On old wood and bark, Adirondack Mountains, New York. Dr. George A. Rex. Sporangium .5-.8 mm. in diameter, the stipe equal in height to the diameter of the sporangium.

4. Trichia erecta, Rex. Sporangium globose to globose-turbinate, stipitate; the wall of both sporangium and stipe with a rough outer layer of brown scales and granules, which, on the upper surface of the sporangium, soon breaks up into [Pg 37] irregular patches. Stipes long, erect, usually simple, rarely fasciculate and connate. Mass of capillitium and spores, bright yellow; elaters simple, cylindric, 4 mic. in thickness, ending in smooth points, 4–6 mic. long; spirals four, often united by intervening branches, covered with numerous irregular spinules. Spores globose, minutely warted, 12–14 mic. in diameter.

Growing on old wood and bark, Adirondack Mountains, New York, Dr. Geo. A. Rex. Sporangium .5-.8 mm. in diameter, the stipe about 1 mm. in height. This Trichia is conspicuous by the checkering or areolation of the upper surface in the mature sporangia, affording a sharp contrast between the brown patches and the yellow bands.

§2. Chrysophidia. Sporangia globose, obovoid or somewhat irregular, sessile, rarely with a short stipe, usually closely crowded. Spores globose, the surface minutely warted.

a. Elaters perfectly smooth.

5. Trichia varia, Pers. Sporangia globose, obovoid or somewhat irregular, gregarious and scattered or crowded, yellowish, ochraceous or olivaceous, sessile, or with a very short brown or blackish stipe. Mass of capillitium and spores yellow; elaters long, simple or sometimes branched a time or two, 4–5 mic. in thickness, ending in a smooth tapering point, 8–12 mic. long; spirals only two, smooth, very prominent in places, causing the elater to appear notched. Spores globose, oval or somewhat irregular, minutely warted, 10–14 mic. in diameter.

Growing in patches on old wood; a very common species. Sporangium .6-.8 mm. in diameter, or when irregular sometimes elongated to 1 mm. or more. Extremely variable as to the form of the sporangium, but readily recognized by its elaters.

6. Trichia Andersoni, Rex. Sporangia globose or obovoid, sessile, gregarious, closely crowded, or sometimes scattered, the wall thickened with minute scales, in color brownish-ochre or olivaceous. Mass of capillitium and spores yellow; elaters long, simple, 3-4 mic. in thickness, ending in [Pg 38] a very long flexuous point, 14-18 mic. in length; spirals three or four, winding evenly and closely, perfectly smooth. Spores globose, minutely warted, 10-12 mic. in diameter.

Growing on the inside of bark of Acer. Sporangium .4-.5 mm. in diameter. The capillitium is deep orange and the spores olivaceous, but this difference in shade of color between spores and capillitium occurs in other species. *Trichia advenula*, Mass., is a closely related species, the swellings in the elaters having no specific value.

7. Trichia inconspicua, Rost. Sporangia very small, subglobose, sessile, collected together in clusters, or scattered, without any hypothallus; the wall brown, smooth and shining. Mass of capillitium and spores yellow; elaters long, simple, cylindric, 3-4 mic. in thickness, ending in smooth tapering points, 6-7 mic. in length; spirals three or four, close, not prominent, perfectly smooth. Spores globose, minutely warted, 10-12 mic. in diameter.

Growing on bark of Platanus, etc. New York, *Peck*; Iowa, *McBride*. The sporangia spherical or reniform and very small.

b. Elaters spinulose.

8. Trichia Iowensis, McBride. Sporangia subglobose, sessile, gregarious, scattered, or sometimes close and confluent; the wall thickened with minute scales, reddish-brown in color. Mass of capillitium and spores yellow; elaters quite variable, usually very long, but sometimes very short, simple, rarely branched, the thickness unequal, 3-4 mic. in the same elater, with occasional thicker swellings, bearing numerous scattered spines, usually about as long as the thickness of the elater, but sometimes much longer, those at the ends being similar; spirals three or four, fine and close, in places

nearly obsolete. Spores globose, or more or less irregular, minutely warted, 9–11 mic. in diameter.

Growing on old bark of Populus; Iowa, McBride. Sporangia .4–.5 mm. in diameter. This is a very curious species of Trichia; it suggests *Ophiotheca Wrightii*, but the elaters are short and simple, and there is no question as to the spirals upon them. I could find no branched elaters in my specimen. [Pg 39]

9. Trichia scabra, Rost. Sporangia globose or somewhat irregular, sessile and closely crowded on a well-developed hypothallus; the wall thin, gold-yellow or orange to yellow-brown in color, smooth and shining. Mass of capillitium and spores orange or golden-yellow; elaters long, simple, 4–5 mic. in thickness, ending in a smooth tapering point, 5–8 mic. in length; spirals three or four, covered with numerous short acute spinules. Spores globose, minutely warted, 9–11 mic. in diameter. See Plate I, Fig. 23.

Growing on old wood in patches, sometimes several centimeters in extent. Sporangia .6–1 mm. in diameter. "The papillæ, which cover the spore, show, when highly magnified, a distinct net-like pattern," *McBride*. The elaters of this species are subject to much irregularity in the way of abnormal swellings, duplicating the spines at the apex, etc.; the spinules are sometimes quite obsolete on some or all of the elaters of a sporangium.

§3. Goniospora, Fr. Sporangia obovoid to oblong, sessile and closely crowded on a well-developed common hypothallus. Spores with thick ridges upon the surface, which are combined into a more or less incomplete network of polygonal meshes.

The ridges of the epispore are 1–2 mic. in height, and do not present to the view more than two or three perfect polygons on a hemisphere of the spores; more often the reticulation is imperfect, the ridges being interrupted and defective. When highly magnified these ridges are seen to be "perforated through their thickness with one, two or three rows, or with clusters of cylindrical openings or pits, or are sculptured into intricate plexuses of minute reticulations with quadrilateral interspaces."

10. Trichia affinis, DeB. Sporangia obovoid to oblong, sessile and closely crowded on a common hypothallus; the wall thin, golden-

yellow to tawny or brownish-yellow, smooth and shining. Mass of capillitium and spores golden to tawny-yellow; elaters long, simple, 4–5 mic. in thickness, ending in a smooth tapering point, 6–10 mic. in length; spirals four, usually spinulose, rarely smooth. Spores angularly or irregularly globose, 10–12 mic. in diameter. [Pg 40]

Growing on old wood and bark in small patches of a few millimeters to a centimeter or more in extent. Sporangia .6-.8 mm. in height by .4-.5 mm. in diameter. *Trichia Jackii*, Rost., is included in this species.

11. Trichia chrysosperma, Bull. Sporangia oblong-obovoid to cylindric, sessile and closely crowded on a well-developed hypothallus; the wall thin, pale citron to olive-yellow, smooth and shining. Mass of capillitium and spores, golden to ochre-yellow; elaters long, simple, 6–8 mic. in thickness, ending in a smooth tapering point, 3–7 mic. in length; spirals four or five, usually smooth, rarely spinulose. Spores angularly or irregularly globose, 12–14 mic. in diameter.

Growing on old wood, in small patches, one to several centimeters in extent. Sporangia 1–2 mm. in height and .5-.6 mm. in diameter. This is readily distinguished from *Trichia affinis* by the larger and differently colored sporangia.

IV. OLIGONEMA, Rost. Sporangia subglobose, more or less irregular, sessile and closely crowded, often in heaps, one upon another, the wall thin, smooth and shining; hypothallus none. Capillitium scanty, composed of elaters habitually irregular and abnormal, intermingled with the spores; elaters simple or sometimes branched, commonly very short, but varying greatly in length, even in the same sporangium; the surface marked with faint spirals, with a few annular ridges, minutely punctulate or altogether smooth. Spores globose, yellow.

The species of this genus are to be regarded as degenerate Trichias. Of course, the abnormality is exhibited most markedly by the elaters; nevertheless, the sporangia of some of the species have a peculiar habit of heaping themselves upon each other.

A. *Surface of the spores reticulate.*

a. *Elaters with projecting rings.*

1. Oligonema nitens, Lib. Sporangia subglobose, irregular, sessile, closely crowded and heaped upon each other, the wall thin, yellow, smooth and shining. Mass of capillitium [Pg 41] and spores yellow; elaters simple or sometimes branched, 3-4 mic. in thickness, with a few distant projecting rings, the surface smooth between, or with very faint spirals, the extremities obtuse, or sometimes with a minute apiculus. Spores angularly or irregularly globose, the surface reticulate, 11-14 mic. in diameter.

Growing in small patches on and within rotten wood. Sporangia .4-.5 mm. in diameter; the elaters variable, some with as many as a dozen projecting rings, some with but a few or nearly smooth. *Trichia nitens*, Libert.

2. Oligonema pusilla, Schr. Sporangia subglobose, irregular, sessile, scattered or collected together in heaps; the wall thin, yellow, smooth and shining. Mass of capillitium and spores yellow; elaters simple or sometimes branched, 4 mic. in thickness, sometimes with thicker inflated portions, the surface marked with low faint spirals or perfectly smooth; the extremities rounded and usually terminating in a smooth point, 3-5 mic. in length — this point either curved, bent to one side or turned back, and twisted around the extremity as a ring. Spores angularly or irregularly globose, the surface reticulate, 11-14 mic. in diameter.

Growing in small clusters in rotten wood. Sporangia .3-.5 mm. in diameter; the elaters variable in length, scarcely exceeding 100 mic. and often much shorter. *Trichia pusilla*, Schroeter.

b. *Elaters with no projecting rings.*

3. Oligonema flavidum, Peck. Sporangia obovoid to oblong, sessile, closely crowded and irregular from mutual pressure; the wall thin, yellow, shining, punctulate or minutely granulose. Mass of spores and capillitium yellow; elaters simple or sometimes branched, 3-4 mic. in thickness, sometimes with thicker inflated portions; the surface punctulate or minutely warted, occasionally marked with very faint spirals; the extremities usually rounded and obtuse, sometimes acute, and rarely with a minute apiculus. Spores

angularly or irregularly globose, the surface reticulate, 11–14 mic. in diameter. See Plate I, Fig. 24. [Pg 42]

Growing in dense patches on old wood and mosses. Sporangia .4-.6 mm. in diameter, and reaching 1 mm. in height, the elaters usually rather long, sometimes quite long and branched.

4. Oligonema brevifila, Peck. Sporangia subglobose, irregular, sessile, crowded, forming clusters or effused patches; the wall thin, yellow, densely granulose and venulose. Mass of capillitium and spores ochre-yellow; elaters simple or sometimes branched, often very short and fusiform, when elongated having long tapering extremities, sometimes with irregular swollen portions; the surface minutely granulose and rugulose, here and there a few spinules, occasionally with indistinct spirals. Spores angularly or irregularly globose, the surface reticulate, 11–12 mic. in diameter.

Growing on old wood and mosses. Sporangia .4-.5 mic. in diameter, the elaters varying greatly in length, some not more than 20 or 30 mic. long, others more than 100 mic. in length.

B. *Spores minutely warted.*

5. Oligonema fulvum, Morgan n. sp. Sporangia rather large, subglobose, sessile, closely crowded and more or less irregular; the wall tawny yellow, very thin and fragile, smooth, shining and iridescent. Mass of capillitium and spores tawny yellow; elaters simple or sometimes branched, mostly very short, 4 mic. in thickness, sometimes with thicker swollen portions; the surface marked with low smooth spirals, in places faint and obsolete; the extremities rounded and obtuse, usually with a very minute apiculus, 1–3 mic. in length. Spores globose, minutely warted, 10–13 mic. in diameter.

Growing on an old effused Sphæria. Sporangia .6-.8 mm. in diameter, the elaters mostly 40–80 mic. in length, rarely much longer and sometimes shorter; the longer elaters and those that are branched often arise from confluence of the shorter ones.

EXPLANATION OF PLATE I.

- Fig. 13. — Perichæna depressa, Lib.
- Fig. 14. — Ophiotheca Wrightii, B. & C.

- Fig. 15.—Lachnobolus globosus, Schw.
- Fig. 16.—Arcyria Cookei, Massee.
- Fig. 17.—Arcyria minor, Schw.
- Fig. 18.—Heterotrichia Gabriellæ, Massee. (After Massee.)
- Fig. 19.—Hemiarcyria plumosa, Morgan.
- Fig. 20.—Hemiarcyria funalis, Morgan.
- Fig. 21.—Calonema aureum, Morgan.
- Fig. 22.—Trichia fallax, Pers.
- Fig. 23.—Trichia scabra, Rost.
- Fig. 24.—Oligonema flavidum, Peck.

Note.—Each figure exhibits the sporangium as it appears magnified about 100 diameters, and the capillitium and spores magnified about 500 diameters.

The Journal of the Cin. Soc. Natural History
Vol. XVI. Plate I.

[Pg 43]

THE MYXOMYCETES OF THE MIAMI VALLEY, OHIO.

By A. P. Morgan.

Third Paper.

(Read February 6, 1894.)

Order VI. STEMONITACEÆ.

Sporangia globose or ovoid to oblong and cylindrical, stipitate; the wall very thin and fragile, soon disappearing. Stipe tapering upward and continued within the sporangium as a more or less elongated columella. Capillitium of slender brown threads, arising from numerous points of the columella, repeatedly branching and usually anastomosing to form a network, persistent and rigidly preserving the outline of the sporangium. Spores globose, brown or violaceous.

This order is readily distinguished by the brown persistent capillitium, arising from a lengthened columella, and rigidly maintaining the form of the sporangium.

Table of Genera of Stemonitaceæ.

A. Stipe and columella brown or black.

a. The columella scarcely reaching the center of the sporangium.

- 1. Clastoderma. Threads of the capillitium forking several times, but not combined into a network. [Pg 44]
- 2. Lamproderma. Threads of the capillitium branching and anastomosing to form a network.

b. The columella extending beyond the center of the sporangium.

- 3. Comatricha. Threads of the capillitium forming only an interior network, attaining the wall by numerous more or less elongated free extremities.

- 4. Stemonitis. Threads of the capillitium forming an interior network of large meshes and a superficial network of smaller meshes.
- 5. Enerthenema. Threads of the capillitium pendent from a discoid membrane at the apex of the columella.

B. Stipe and columella white or yellowish.

- 6. Diachaea. Threads of the capillitium branching and anastomosing to form a network.

I. CLASTODERMA, Blytt. Sporangium regular, globose, stipitate; the wall very thin and fragile. Stipe elongated, tapering upward, entering the sporangium as a very short or nearly obsolete columella. Capillitium arising by a few branches from the apex of the columella, these branches forking several times at a sharp angle, but not combined into a network, the ultimate branchlets long and free, or only connected together at their tips by persistent fragments of the sporangial wall. Spores globose, violaceous.

The claim of this genus to be distinguished from Lamproderma must rest upon the fact that the branchlets of the capillitium do not anastomose and form a network. It is the same as the genus Orthotricha of Wingate.

1. Clastoderma De Baryanum, Blytt. Sporangium very small, globose; the wall early disappearing, except the minute [Pg 45] fragments which persist at the extremities of the capillitium, and a narrow collar at the base of the columella. Stipe very long, thick and brown below, tapering upward to a pellucid oblong swelling, thence abruptly narrowed to the apex; the columella extremely short, capillitium of very slender pale-brown semi-pellucid threads, divergently forking, the ultimate branchlets often joined 2–4 together at their tips by fragments of the sporangial wall. Spores globose, even, violaceous, 8–9 mic. in diameter. See Plate XI, Fig. 25.

Growing in rather a scattered way on old rotten wood. Sporangium .20–.25 mm. in diameter, the stipe .7–1.3 mm. long. *Orthotricha microcephala*, Wingate. Blytt's species was found in Norway, Wingate's in Pennsylvania; I have met with it several times in this locality. It is possibly more common than it appears, as by reason of the difficulty of seeing the minute sporangium it is passed by as some mold. Blytt's spore measurements are 9.5-11 mic.; in some specimens I have seen a few spores of this size, but they are abnormal.

II. LAMPRODERMA, Rost. Sporangia regular, globose, stipitate; the wall thin and fragile, rugulose, shining with metallic tints, breaking up irregularly and gradually falling away. Stipe more or less elongated, smooth, brown or black in color, arising from a hypothallus, tapering upward and entering the sporangium as a short columella scarcely reaching the center. Capillitium of numerous threads radiating from the columella, usually forking several times and combined into a net by lateral anastomosing branchlets. Spores globose, brown or violaceous.

Lamproderma is distinguished by the shining metallic tints of the sporangial wall, and by the short columella scarcely reaching half the height of the sporangium.

1. Lamproderma physaroides, A. & S. Sporangium globose; the wall with a silvery metallic luster, at length breaking up and falling away. Stipe long, slender, brown or blackish, arising from a small circular hypothallus; columella clavate, obtuse, not reaching the center of the sporangium. Cap [Pg 46] illitium of brownish-violet threads, arising from the upper part of the columella; these branch repeatedly at a sharp angle, form an intricate network of elongated meshes, terminating at the wall in numerous short free branchlets. Spores globose, minutely warted, bright brown, 12-14 mic. in diameter.

Growing on old wood, moss, etc., New York, *Chas. H. Peck*. Distinguished by the pale silvery sporangial wall and the clear brown spores.

2. Lamproderma arcyrionema, Rost. Sporangium small, globose; the wall dark bronze, with a silvery sheen when loosened from the spores, soon breaking into scales and falling away. Stipe long and slender, smooth, shining and black, rising from a thin hypothallus; the columella short cylindric, variable in length, but not attaining the center of the sporangium. Capillitium arising by division of the apex of the columella into several primary branches; these immediately separate into numerous slender flexuous brown threads,

which unite and form a dense network of small arcuate meshes, the ultimate branchlets not free. Spores globose, even, violaceous, 6–7 mic. in diameter. See plate XI, Fig. 26.

Growing on old wood of Juglans and Carya. Sporangium .3–.5 mm. in diameter, the stipe three or four times as long. The columella is somewhat variable, it sometimes forks or divides immediately on entering the sporangium, at other times it is longer and cylindric, with more slender primary branches. The meshes of the capillitium resemble those of Arcyria, whence the name. This is the *Stemonitis physaroides*, A. & S. var. *subœneus* of Lea's Catalogue.

3. Lamproderma violaceum, Fr. Sporangium depressed-globose, convex above and more or less flattened and umbilicate beneath; the wall shining with steel or violet, blue and purple tints, deciduous. Stipe short, stout, brown or blackish in color, arising from a thin, brown, common hypothallus; columella cylindric, or tapering slightly to an obtuse apex, attaining the center of the sporangium. Capillitium of numerous slender threads, radiating from the upper part of the columella; these threads are brown below, with a vari [Pg 47] able outer portion colorless; they branch a few times and form an interior network of elongated meshes, outwardly arching and freely anastomosing they give rise to an external network of small irregular meshes, they then attain the wall by innumerable short, simple, or forked free branchlets. Spores globose, minutely spinulose, violaceous, 9–11 mic. in diameter. See plate XI, Fig 27.

Growing on old wood, mosses, etc., late in Autumn. Sporangium .5–.8 mm. in diameter, the stipe about the same length. The capillitium is sometimes most of it colorless and flaccid; sometimes it is all brown and rigid except the minute free extremities.

4. Lamproderma arcyrioides, Somm. Sporangium globose or ellipsoid, and somewhat elongated; the wall with tints of violet, purple, and blue, deciduous. Stipe usually short, or sometimes nearly obsolete, brown or blackish in color, arising from a strongly-developed hypothallus; the columella cylindric or slightly tapering upward, and obtuse, reaching nearly to the center of the sporangium. Capillitium of numerous pale-brown threads, radiating from the apex of the columella; these fork directly from the base, are bent and flexuous, and are combined into a dense, intricate net, with

abundant free extremities. Spores globose, spinulose, violaceous, 13–16 mic. in diameter.

Growing on old leaves, wood, etc. Sporangium .5-.8 mm. in diameter, the stipe variable in length from very short to 1 mm. long or beyond. *Lamproderma columbinum*, Pers. is a doubtful species, the forms of that name being easily distributed between the present species and *L. physaroides*.

5. Lamproderma scintillans, B. & Br. Sporangium globose; the wall shining with colors of blue, purple, and bronze, deciduous. Stipe long, slender, smooth, and shining, brown or blackish, rising from a thin, brown, common hypothallus; columella cylindric or slightly tapering to the obtuse apex, not reaching the center of the sporangium. Capillitium of numerous brown threads, originating about the apex of the columella; these fork several times, with few anastomosing branchlets, and terminate at the wall in long, free extremities. Spores globose, minutely warted, violaceous, 7–9 mic. in diameter. See Plate XI, Fig. 28. [Pg 48]

Growing on old leaves, moss, etc., in early Spring. Sporangium .3-.5 mm. in diameter, the stipe from once to twice as long. This is *Lamproderma irideum* of Massee's Monograph. I am indebted to Arthur Lister, Esq., of London, for the identification of my specimens with *Stemonitis scintillans*, B. & Br., and with *Lamproderma irideum*, Cke.

III. COMATRICHA, Preuss. Sporangia various in shape, from globose or ovoid to oblong and cylindric, stipitate; the wall very thin and fugacious. Stipe more or less elongated, smooth and black, arising from a common hypothallus, tapering upward, entering the sporangium and prolonged nearly or quite to the apex as a columella. Capillitium arising from numerous points of the columella throughout its entire length; the threads immediately branching and anastomosing to form an interior network, attaining the wall by numerous more or less elongated free extremities. Spores globose, brown or violaceous.

This genus is not sharply limited from Stemonitis. The species with very short free ends, and consequently with superficial meshes approximate to the wall, are near the form of Stemonitis. But it may be observed that in these species, the meshes of the capillitium become smaller gradually outward, the sides of the superficial meshes are arched away from the wall, and they are in contact with it only by the free extremities.

§1. Typhoides. Threads of the capillitium repeatedly branching and anastomosing, to form a dense network of small meshes, with innumerable short, free extremities.

1. Comatricha typhina, Roth. Sporangia short, erect or a little curved, cylindric or usually narrowing slightly upward, the base quite blunt, the apex more rounded, growing together on a thin hypothallus. Stipe and columella brown or blackish, tapering upward and vanishing near the apex of the sporangium, the stipe much shorter than the columella. Capillitium of slender flexuous tawny-brown threads; these branch repeatedly, forming an intricate network of small irregular meshes, ending in very short free [Pg 49] extremities. Spores globose, violaceous, very minutely warted, 6–8 mic. in diameter.

Growing on old wood, mosses, etc. Sporangium with the stipe 2–4 mm. in height, the stipe much the shorter, the sporangium .35–.40 mm. in thickness. *Stemonitis typhoides*, Fries, S. M.

2. Comatricha æqualis, Pk. Sporangia usually more or less inclined or curved and nodding, cylindric, obtuse at each end, growing together on a thin hypothallus. Stipe and columella slender, smooth, black, extending nearly or quite to the apex of the sporangium, the stipe longer than the columella. Capillitium of very slender flexuous tawny-brown threads; these branch repeatedly, forming an intricate network of small irregular meshes, ending in very short free extremities. Spores globose, minutely warted, dark violaceous, 7–9 mic. in diameter.

Growing on old wood. Sporangium 1.5–3 mm. in height by .35–.40 mm. in thickness, the stipe usually about the same length as the sporangium, but sometimes nearly twice as long. The capillitium is rather looser than in *C. typhina*, whence the drooping habit. Peck, Thirty-first Report, p. 42.

3. Comatricha nigra, Pers. Sporangia globose or ovoid to ellipsoid or oblong, erect or sometimes inclined or even nodding. The stipe very long, smooth and black, tapering upward, expanding at the

base into a small circular hypothallus; the columella short, reaching from one-half to three-fourths the height of the sporangium. Capillitium of slender flexuous brown threads, which branch repeatedly, forming a dense intricate network of small meshes, ending in very short free extremities. Spores globose, even, dark violaceous, 8-10 mic. in diameter.

Growing on old wood, leaves, etc. Sporangium .5-1.5 mm. in height, .5-.8 mm. in diameter, the stipe 1.5-3 mm. long or sometimes considerably longer. This species seems to be rare in this country. I have preferred the name adopted by Schroeter to Rostafinski's *Comatricha Friesiana*.

4. Comatricha Ellisii, Morgan, n. sp. Sporangia short, erect, oval or ovoid to oblong. Stipe and columella erect, brown and smooth, rising from a thin pallid hypothallus, [Pg 50] tapering upward and vanishing into the capillitium toward the apex of the sporangium, the stipe usually longer than the columella. Capillitium of slender pale brown threads; these branch several times with lateral anastomosing branchlets, forming a rather open network of small meshes, ending with very short free extremities. Spores globose, even, pale ochraceous, 6-7 mic. in diameter. See Plate XI, Fig. 29.

Growing on old pine wood. Sporangium .3-.6 mm. in height by .3-.5 mm. in width, the stipe usually a little longer than the sporangium. This elegant little species I have from Mr. J. B. Ellis, of Newfield, N. J. It is said to be mingled in some of the specimens with *Lamproderma Ellisiana*, Cke.

§2. Larvella. Threads of the capillitium branching a few times and anastomosing to form a network of large meshes, attaining the wall by numerous long, free extremities.

5. Comatricha crypta, Schw. Sporangia cylindric, bent or flexuous and more or less inclined, growing close together on a conspicuous purplish-brown hypothallus. Stipe and columella smooth and black, tapering upward and reaching the apex of the sporangium, the columella bent and flexuous or spirally twisted, about as long as the stipe. Capillitium composed of irregular, bent and uneven threads, which are brown below, becoming colorless outwardly; the threads branch a few times, forming a network of large irregular meshes, sometimes much defective; the free extremities irregular and une-

qual, simple or branched. Spores globose, brown, minutely warted, 7-9 mic. in diameter. See Plate XI, Fig. 30.

Growing out of fissures of the bark and wood of Hickory, Acer, etc. Sporangium with the stipe 4-7 mm. in height, the stipe a little shorter, or sometimes much longer than the sporangium, the latter .25-.30 mm. in thickness. The exterior colorless portion of the capillitium is exceedingly delicate, easily breaking away and leaving the capillitium quite irregular and defective. *Stemonitis crypta*, Schweinitz's N. A. Fungi, 2351. *Comatricha irregularis*, Rex, is the same thing.

6. Comatricha cæspitosa, Sturgis. Sporangia short, clavate, densely crowded or cæspitose upon a delicate hypothallus; the wall subpersistent, silvery, shining with tints of [Pg 51] purple and blue. Stipe very short or nearly obsolete, the columella rising to two-thirds or three-fourths the height of the sporangium. Capillitium of slender dark-brown threads, which branch and anastomose quite irregularly, forming a network of intermingled large and small meshes, ending in long, tapering, free extremities. Spores globose, minutely spinulose, dark violaceous, 10-12 mic. in diameter.

Growing on moss and lichens, at Wood's Holl, Massachusetts. Sporangium 1-1.5 mm. in height, the stipe very short or sometimes apparently wanting. I am indebted to Dr. W. C. Sturgis, of New Haven, Conn., for a specimen of this unique species.

7. Comatricha longa, Peck. Sporangia very slender and much elongated, tapering gradually upward, weak and prostrate or pendulous, growing close together on a well-developed purplish-black hypothallus. Stipe and columella capillary, smooth and black, reaching to the apex of the sporangium or often vanishing in the network far below it, the stipe very short, the columella long and flexible. Capillitium of long, slender, dark-brown threads; these are reticulately connected near the base, forming a network of large irregular meshes in a series along the columella; outwardly they are terminated by very long free branchlets, which vary from simple to two or three times forked or branched. Spores globose, minutely warted, dark brown, 8-10 mic. in diameter. See Plate XI, Fig. 31.

Growing on old wood and bark of Elm, Willow, etc., in Autumn. Sporangium with the stipe 15-40 mm. in length, the stipe 3-8 mm.

long, the sporangium .25-.40 mm. in thickness. This is the most characteristic species of the genus, being farthest removed from Stemonitis.

8. Comatricha flaccida, Lister. Sporangia growing closely crowded together and more or less confluent, on a purplish-brown hypothallus, the walls fugacious. Columellas rising simply from the common hypothallus, or sometimes grown together below and then apparently branching, running through to the apex, and there often confluent with each other, or joined together by portions of membrane. Capillitium of slender brown threads, which branch and anasto [Pg 52] mose very irregularly, forming a ragged network with large irregular meshes, and long free extremities; the capillitium of adjoining columellas being much entangled, and often confluent or grown together. Spores globose, very minutely warted, brown, 7–9 mic. in diameter.

Growing on old wood and bark of Oak, Willow, etc. The component sporangia 5–10 mm. in length. The early appearance is much like that of species of Stemonitis, but the mature stage is a great mass of spores with scanty capillitium, as in Reticularia; the columellas, however, are genuine and not adjacent portions of wall grown together. Arthur Lister calls this *Stemonitis splendens*, var. *flaccida*.

IV. STEMONITIS, Gled. Sporangia subcylindric, elongated, stipitate, standing close together on a well-developed common hypothallus, the wall very thin and evanescent. Stipe brown or black, smooth and shining, tapering upward, entering the sporangium and prolonged nearly to the apex as a slender columella, the stipe shorter than the columella. Capillitium arising from numerous points of the columella throughout its entire length; the threads immediately branch and anastomose to form an interior network of large meshes, they then spread out next the wall of the sporangium into a superficial network of smaller meshes. Spores globose, brown or violaceous.

In this genus there are two distinctly differentiated series in the capillitium, the one an interior supporting network of large meshes, the other a superficial network of smaller meshes; sometimes the superficial network disappears or is wanting toward the upper part of the capillitium, there is then an approach to Comatricha. Very minute scattered branchlets usually connect the superficial network with the wall of the sporangium.

§1. Dictynna. Threads of the capillitium arising from numerous points of the columella, immediately branching several times and anastomosing to form the interior network of large meshes; the superficial network consisting of small irregular and unequal meshes, varying from smaller than the spores to two or three times their diameter. [Pg 53]

1. Stemonitis fusca, Roth. Sporangia elongated, subcylindric, tapering and obtuse at the apex, tapering gradually downward, growing closely crowded together on a strongly-developed brown hypothallus. Stipe and columella smooth and black, tapering gradually upward and disappearing near the apex of the sporangium, the stipe shorter than the columella. Capillitium of slender brown or blackish threads, which immediately branch and anastomose, forming a dense interior network of large irregular meshes, the ultimate branchlets of which support a superficial network of small polygonal meshes. Spores globose, dark violaceous, the surface minutely warted, the warts with a reticulate arrangement, 7–9 mic. in diameter.

Growing on old wood, bark, leaves, etc.; common everywhere. Sporangium with the stipe 6–15 mm. in height, the sporangium .3–.4 mm. in thickness, the stipe variable in length, but always shorter than the sporangium. The meshes of the superficial net vary in size in the same sporangium, being usually 5–25 mic. in width, but sometimes they are larger, ranging from 10–40 mic. in extent. The name *Stemonitis maxima* was given by Schweinitz to some unusually large specimens which grew on a Polyporus. *Stemonitis dictyospora* of Rostafinski's monograph, with spores 12 mic. in diameter, is said to occur in South Carolina; I have seen no specimens.

2. Stemonitis tenerrima, B. & C. Sporangia small, subcylindric, tapering and obtuse at the apex, tapering gradually downward, growing close together on a thin brown hypothallus. Stipe and columella black and smooth, tapering gradually upward and vanishing toward the apex of the sporangium, the stipe shorter than the columella. Capillitium of very slender pale violet threads, which branch

and anastomose to form a dense interior network of large irregular meshes, and then spread out into a superficial network of small polygonal meshes. Spores globose, even, pale brownish-violet, 6–8 mic. in diameter. See Plate XI, Fig. 32.

Growing on old wood, mosses, etc. Sporangium with the stipe 5–9 mm. in height, the sporangium .2–.3 mm. in thickness, the stipe variable in length, but always shorter than the sporangium. The meshes of the superficial network varying [Pg 54] usually from 3–15 mic. in width, but sometimes larger from 8–25 mic. The species grows scantily in this region, but I have elegant specimens from Alabama, sent me by Prof. Geo. F. Atkinson.

3. Stemonitis microspora, Lister. Plasmodium white. Sporangia elongated, subcylindric, tapering and obtuse at the apex, tapering gradually downward, growing closely crowded together on a strongly-developed brown hypothallus. Stipe and columella brown and smooth, tapering gradually upward and reaching nearly to the apex of the sporangium, the stipe shorter than the columella. Capillitium of slender tawny-brown threads; the primary branches simple or only branched above, or with a few lateral anastomosing branchlets, forming a rather loose network of large irregular meshes; these support a superficial network of very small polygonal meshes. Spores globose, even, tawny-brown, 5–6 mic. in diameter.

Growing on old wood, bark, leaves, etc.; very common in this region. Sporangium with the stipe 7–15 mm. in height, the sporangium .3–.4 mm. in thickness, the stipe shorter than the sporangium. Meshes of the superficial network 4–20 mic. in width. I am indebted to Arthur Lister, Esq., of London, for pointing out to me the difference between this species and the *Stemonitis ferruginea* of Fries and Rostafinski.

4. Stemonitis ferruginea, Ehr. Plasmodium lemon-yellow. Sporangia subcylindric, the apex obtuse, growing closely crowded together on a thin, brown hypothallus. Stipe and columella brown and smooth, tapering gradually upward and vanishing beneath the apex of the sporangium, the stipe much shorter than the columella. Capillitium of slender, tawny-brown threads, which immediately branch and anastomose, forming a dense interior network of large irregular meshes, supporting a superficial network of small polygo-

nal meshes. Spores globose, very minutely warted, tawny-brown in color, 7-9 mic. in diameter.

Growing on old wood, leaves, grasses, etc. Sporangium with the stipe 4-10 mm. in height, the sporangium .3-.4 mm. in thickness, the stipe much shorter than the sporangium. The meshes of the superficial network varying from 6-30 mic. or sometimes from 10-40 mic. in width, according to the [Pg 55] specimen. The species is certainly rare in this country, and my description is drawn up from British specimens. But I am unable to distinguish authentic specimens of *Stemonitis herbatica*, Peck, from these British specimens.

§2. Megalodictys. Threads of the capillitium arising from rather distant points of the columella, branching and anastomosing but a few times, thus forming an interior network of very large meshes; the superficial network consisting of large irregular meshes, varying from three or four to many times the diameter of the spores.

5. Stemonitis Webberi, Rex. Sporangia subcylindric, the apex obtuse, tapering gradually downward, growing close together on a common hypothallus. Stipe and columella black and smooth, the stipe very short, the columella extending nearly or quite to the apex of the sporangium, the upper part usually flexuous. Capillitium composed of slender, flexuous brown threads; these immediately branch and anastomose several times, forming an interior network of very large meshes; the superficial network consisting of large irregular meshes, sometimes much elongated. Spores globose, very minutely warted, brown, 7-9 mic. in diameter. See Plate XI, Fig. 34.

Growing on old wood, bark, etc. Sporangium with the stipe 5-10 mm. in height, the stipe 1-2 mm. in length, the sporangium .3-.4 mm. in thickness; meshes of the superficial net of the capillitium 40-100-150 mic. in extent. This is a much smaller species than *Stemonitis splendens*, and the structure of the interior network of the capillitium is entirely different.

6. Stemonitis splendens, Rost. Sporangia linear-cylindric, obtuse at the apex, growing close together on a conspicuous hypothallus. Stipe and columella black and shining, the stipe very short, the columella reaching nearly or quite to the apex of the sporangium, often flexuous above. Capillitium composed of brown threads, variable in thickness, often with membranaceous expansions; the primary

branches some of them simple or only branched above, others with a few anastomosing branchlets, forming an interior network of extremely large meshes; the superficial network consisting [Pg 56] of large, irregular, roundish or polygonal meshes. Spores globose, very minutely warted, brown, 7–9 mic. in diameter. See Plate XI, Fig. 33.

Growing on old wood. Sporangium with the stipe 15–25 mm. in height, the stipe 4–6 mm. in length, the sporangium about .4 mm. in thickness; the meshes of the superficial network of the capillitium 25–50–80 mic. or sometimes as much as 100 mic. in extent. This is *Stemonitis Morgani*, Peck.

V. ENERTHENEMA, Bowm. Sporangium regular, globose, stipitate; the wall thin and fragile, fugacious. Stipe stout, thick, tapering upward, entering the sporangium and prolonged to its apex, there expanding into a discoid membrane. Capillitium originating from the lower surface of the apical disk of the columella; the threads branched a few times and hanging downward, their extremities free. Spores globose, violaceous.

A well-marked genus, by reason of the peculiar origin of the capillitium.

1. Enerthenema papillatum, Pers. Sporangium globose, stipitate; the wall brown or blackish, soon disappearing. Stipe black, rugulose, thick below, tapering above into the slender columella, which, at its apex, expands into a thin membranaceous disk. Capillitium of long brown threads suspended from the apical disk, the threads branched a few times, occasionally anastomosing by a short, transverse branchlet, the free ends often forked. Spores globose, very minutely warted, violaceous, 10–12 mic. in diameter. See Plate XI, Fig. 35.

Growing on old wood. Stipe and columella .8–1.2 mm. in height. The species seems to be rare in this country, as I have met with it but once myself, and have received only a few specimens from elsewhere.

VI. DIACHÆA, Fr. Sporangia globose to oblong, stipitate, arising from a common hypothallus; the wall thin, rugulose, iridescent with metallic tints, breaking up irregularly and [Pg 57] gradually falling away. Stipe and columella thick, erect, rigid, tapering upward, filled with minute, roundish granules of lime, white or yellowish in color. Capillitium arising from numerous points of the columella, the threads repeatedly branching and anastomosing to form an intricate network, attaining the wall by numerous short free extremities. Spores globose, violaceous.

This genus is scarcely to be distinguished from Lamproderma, except by the white mass of lime which fills the tube of the stipe and columella.

1. Diachæa leucopoda, Bull. Sporangia ovoid-oblong to short cylindric, the base obtuse or slightly umbilicate, the apex more rounded; the wall with bronze, blue, purple, and violet tints, gradually falling away. Stipe short, thick, white, arising from a white, venulose, hypothallus, tapering upward; the columella cylindric or slightly tapering, obtuse, terminating below the apex of the sporangium. Capillitium of slender, flexuous brown threads forming a dense network of rather small meshes. Spores globose, very minutely warted, violaceous, 7–9 mic. in diameter.

Growing on old leaves, sticks, etc., and sometimes running over living plants. Sporangium with the stipe 1–2 mm. in height, the stipe usually much shorter than the sporangium, the latter .4–.5 mm. in thickness. *Diachæa elegans*, Fr.

2. Diachæa splendens, Peck. Sporangia globose, sometimes a little depressed, with the base umbilicate; the wall steel-blue with tints of purple and violet, quite persistent, rupturing irregularly. Stipe short, thick, white, arising from a white, reticulate hypothallus, tapering upward; the columella oblong or short cylindric, extending beyond the center of the sporangium. Capillitium of slender, brown threads, which branch several times and form a loose network of rather large meshes. Spores subglobose, with very large warts, dark violet, 8–10 mic. in diameter.

Growing on old leaves and twigs. Sporangium .4–.6 mm. in diameter, the stipe about the same length. This is a beautiful species.

3. Diachæa Thomasii, Rex. Sporangia globose, or sometimes a little depressed; the wall gold-bronze, with tints of [Pg 58] purple and blue, subpersistent, rupturing irregularly. Stipe thick, dull ochre-yellow in color, variable in length, usually very short and some-

times quite obsolete, arising from an ochre-yellow hypothallus; the columella varying from bluntly-conical to cylindric-clavate, attaining the center of the sporangium. Capillitium of slender, brown threads, radiating from all points of the columella, branching several times and forming a loose network of elongated meshes. Spores globose, minutely warted, violaceous, 11–12 mic. in diameter. See Plate XI, Fig. 36.

Growing on sticks, leaves, etc. Sporangium .5–.7 mm. in diameter, the stipe usually shorter or sometimes wanting. This species has been found only in the mountains of North Carolina. I am indebted to Dr. George A. Rex for my example. In its structure the species is essentially a Lamproderma, but the stipe and columella are stuffed with granules of lime.

EXPLANATION OF PLATE XI.

- Fig. 25. — Sectional view of the capillitium and stipe of Clastoderma De Baryanum, Blytt.
- Fig. 26. — Section through the capillitium, columella and stipe of Lamproderma arcyrionema, Rost.
- Fig. 27. — Perpendicular section through Lamproderma violaceum, Fr.
- Fig. 28. — Perpendicular section through Lamproderma scintillans, Berk.
- Fig. 29. — Section through the capillitium, columella and stipe of Comatricha Ellisii, Morgan.
- Fig. 30. — Sectional view through the capillitium and columella of a portion of Comatricha crypta, Schw.
- Fig. 31. — Sectional view through the columella and capillitium of a portion of Comatricha longa, Peck.
- Fig. 32. — A portion of the capillitium of Stemonitis tenerrima, B. & C. — A sectional view through the columella above and below a view of the superficial network.
- Fig. 33. — A portion of the capillitium of Stemonitis splendens, Rost. — A sectional view through the columella above and below a view of the superficial network.

- Fig. 34.—The capillitium of a very short sporangium of Stemonitis Webberi, Rex; the breadth, however, somewhat exaggerated.
- Fig. 35.—Showing the stipe, columella, apical disk and pendent capillitium of Enerthenema papillatum, Pers.
- Fig. 36.—Perpendicular section through the capillitium, columella, and stipe of Diachæa Thomasii, Rex.

Note.—The figures of the objects are drawn as they appear under a magnifying power of about 100 diameters.

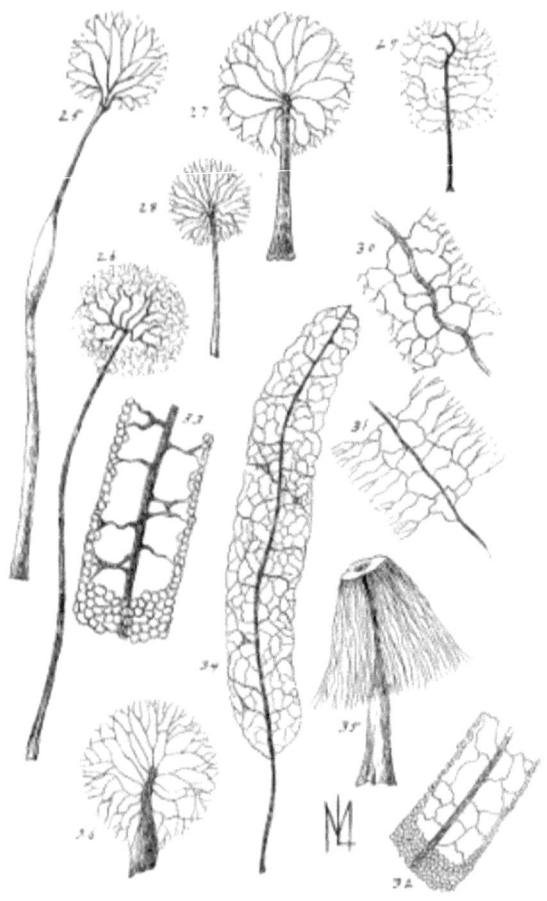

The Journal of the Cin. Soc. Natural History
Vol. XVI. Plate XI.

Order VII. — DIDYMIACEÆ.

Sporangia simple and subglobose, or plasmodiocarp, rarely combined into an æthalium. Wall of the sporangium a thin membrane with an outer layer composed of minute stellate crystals, or of minute roundish granules of lime; these either lie singly upon the surface, or are compacted into a crustaceous coat. Stipe present or often wanting; the columella usually conspicuous and well-developed. Capillitium consisting of very slender, often sinuous threads, which extend from the base of the sporangium or from the columella to the walls, either simple or outwardly branching a few times at a sharp angle, combined into a loose irregular net by a few transverse branchlets, which are situated chiefly at the extremities. Spores globose, violaceous. [Pg 59]

This order is readily distinguished from the Physaraceæ by the absence of lime from the threads of the capillitium.

Table of Genera of Didymiaceæ.

a. The lime on the wall of the sporangium in the form of minute stellate crystals.

- 1. Didymium. Sporangium simple, subglobose and stipitate, the base commonly umbilicate, or sometimes sessile and plasmodiocarp.
- 2. Spumaria. Æthalium composed of numerous elongated irregularly-branched sporangia, closely compacted together and confluent.

b. The lime on the wall of the sporangium consisting of minute roundish granules.

- 3. Diderma. Wall of the sporangium with the outer calcareous layer usually compacted into a smooth continuous crust.

- 4. Lepidoderma. Wall of the sporangium with an outer layer of large scales, consisting of bicarbonate of lime.

I. DIDYMIUM, Schrad. Sporangium simple, subglobose and stipitate, the base commonly umbilicate, or sometimes sessile and plasmodiocarp; the wall a thin membrane with an outer layer of minute stellate crystals of lime. Stipe present or sometimes wanting; the columella mostly conspicuous, sometimes thin or obsolete. Capillitium of very slender threads, straight or often sinuous, stretching from the columella to the wall of the sporangium, simple or outwardly sparingly branched at a sharp angle. Spores globose, violaceous.

Didymium, together with Spumaria, is to be distinguished from all other genera of the Myxomycetes by the covering of stellate crystals, like hoar-frost, upon the outer surface of the sporangium. [Pg 60]

§1. Cionium. Columella prominent, subcentral, globose, obovoid, or turbinate; the threads of the capillitium radiating in all directions to the wall of the sporangium.

A. Sporangium stipitate.

1. Didymium squamulosum, A. & S. Sporangium variable in form and size, small and globose, or large and much depressed, the base usually umbilicate, stipitate, or sometimes sessile, and even plasmodiocarp; the wall very thin and pellucid, with a thin, gray-white layer of stellate crystals of lime, breaking up into subpersistent scales. Stipe short, erect, snow-white, longitudinally furrowed or plicate; the columella central, snow-white, various in shape, globose, obovoid, turbinate, and stipitate or sessile. Capillitium of numerous colorless threads, radiating from the columella and separating outwardly into several branches. Spores globose, very minutely warted, dark violaceous, 8–10 mic. in diameter.

Growing on old wood, leaves, herbaceous stems, etc. Sporangium .4-.6-.8 mm. in diameter, the stipe scarcely longer than the diameter, often much shorter or nearly wanting.

2. Didymium proximum, B. & C. Sporangium globose or depressed-globose, the base more or less umbilicate, stipitate; the wall

very thin and pellucid, with a loose white covering of stellate crystals of lime, the upper part breaking up and falling away. Stipe long, erect, tapering upward, yellow-brown to reddish-brown, expanding at the base into a small hypothallus; the columella central, white, turbinate, or discoid turbinate. Capillitium of slender, colorless threads, radiating from the columella, branching and often anastomosing. Spores globose, even, pale violaceous, 8–10 mic. in diameter. Plate XII, Fig. 37.

Growing on old leaves, sticks, culms, etc. Sporangium .4-.6 mm. in diameter, the stipe two or three times the diameter.

3. Didymium eximium, Peck. Sporangium depressed-globose, the base umbilicate, sometimes very much depressed and also umbilicate above, stipitate; the wall pale ocher or pale yellow, with a thin layer of minute white crystals of [Pg 61] lime, the upper part gradually breaking away. Stipe long, erect, tapering upward, pale yellow-brown, darker below, expanding into a small brown hypothallus; the columella central, large, discoid, or sometimes rough and irregular, pale ochre or yellowish. Capillitium of much-branched colorless threads, radiating upward and downward from the columella. Spores globose, very minutely warted, dark violaceous, 9–11 mic. in diameter. Plate XII, Fig. 38.

Growing on old leaves, sticks, etc. Sporangium .5-.6 mm. in diameter, the stipe about twice the diameter.

4. Didymium microcarpum, Fr. Sporangium small, globose, the base slightly umbilicate, stipitate; the wall a dark-colored membrane, covered with abundant snow-white crystals of lime. Stipe long, slender, erect, delicately striate, yellow-brown to blackish in color, expanded at the base into a small hypothallus; the columella small, globose, sessile or substipitate, pale yellow-brown. Capillitium of pale brown threads, somewhat branched and forming a loose net. Spores globose, very minutely warted, violaceous, 6–7 mic. in diameter.

Growing on old wood, leaves, mosses, etc. Sporangium .4-.5 mm. in diameter, the stipe two or three times as long. The species is more particularly distinguished by its small spores.

5. Didymium minus, Lister. Sporangium depressed-globose, the base umbilicate, stipitate, rarely sessile and plasmodiocarp; the wall a dark-colored membrane with a thin layer of stellate crystals of lime, breaking up gradually and falling away. Stipe erect or sometimes bent at the apex, variable in length, rarely wanting, from pale brown to blackish in color, rising from a small hypothallus; the columella reaching the center, brown or blackish, rough, convex, subglobose or pulvinate, substipitate. Capillitium of slender colorless threads, radiating from the columella and more or less branched outwardly. Spores globose, very minutely warted, violaceous, 8–10 mic. in diameter. Plate XII, Fig. 39.

Growing in vast abundance in Spring on old leaves, bark, wood, etc. Sporangium .4-.6 mm. in diameter, the stipe scarcely longer but usually shorter than the diameter of the [Pg 62] sporangium rarely absent. It is considered by Lister to be a variety of *D. farinaceum*; it differs from this species in its smaller and less-depressed sporangium and in its smaller nearly smooth spores.

B. *Sporangia sessile.*

6. Didymium effusum, Link. Sporangia gregarious or scattered, sessile on a flattened base, convex above, various in shape, subrotund or by confluence effused and venosely creeping; the wall very thin and pellucid, invested with a thin flocculose layer of minute crystals of lime. The columella hemispheric, rugulose, usually snow-white. Capillitium of very slender colorless threads, furnished with numerous minute protuberances, much branched and combined into a dense net. Spores globose, very minutely warted, dark violaceous, 10–11 mic. in diameter.

Growing on old leaves, wood, etc. Sporangium about .5 mm. in diameter or thickness, sometimes confluent and more or less elongated as a plasmodiocarp. This species is reported from the United States, but I have seen no specimens.

7. Didymium physaroides, Pers. Sporangia roundish or hemispheric, more or less irregular and deformed, sessile or with a very short stipe, and closely crowded together upon a strongly-developed common hypothallus; the wall a dark colored membrane, with a thin layer of stellate crystals of lime. The columella large and thick, divided into cells which are filled with irregular

lumps of lime, common to all the sporangia. Capillitium of stout threads, usually simple, only rarely branched, furnished with numerous fusiform swellings. Spores globose, minutely warted, dark violaceous, 12–14 mic. in diameter.

Growing on old wood, bark, moss, etc. Reported from Carolina by Curtis. It is said superficially to resemble somewhat *Physarum didermoides*.

§2. Placentia. Columella basal, much depressed, very thin or quite obsolete, connate with the base of the sporangium; the threads of the capillitium ascending to the wall of the sporangium. [Pg 63]

A. Sporangium stipitate.

8. Didymium farinaceum, Schrad. Sporangium hemispherical, more or less depressed, the base profoundly umbilicate; the wall firm, rugulose, dark-colored and nearly opaque, with a mealy coat of stellate crystals of lime, rupturing irregularly. Stipe variable in length, rigid, erect, black or sometimes rusty-brown, arising from a small hypothallus; the columella broad, hemispherical or pulvinate, black, the lower side connate with the wall of the sporangium. Capillitium of dark-colored sinuous threads, simple or scarcely branched. Spores globose, dark violaceous, minutely warted, 10–13 mic. in diameter. Plate XII, Fig. 40.

Growing on old wood, leaves, mosses, etc. Sporangium .6–.9 mm. in diameter, the stipe about as long as the diameter of the sporangium or sometimes much longer, usually, however, much shorter than the diameter and almost concealed within the umbilicus. My specimens are from Pennsylvania and Alabama. It is readily distinguished from *Didymium minus* by the much larger and more distinctly warted spores.

9. Didymium clavus, A. & S. Sporangium pileate, very much depressed, convex above and concave below, stipitate; the wall a dark-colored membrane, thickly covered with minute white crystals of lime, except the brown concavity underneath, the upper part breaking away, the lower persistent. Stipe short, erect, rugulose, brown or blackish, expanding at the base into a small hypothallus; the columella reduced to a thin layer of minute brown scales upon the base of the sporangium. Capillitium of simple or sparingly-branched

threads, colorless at the extremities and dark-colored between. Spores globose, even, violaceous, 6–8 mic. in diameter.

Growing on old leaves, sticks, herbaceous stems, etc. The sporangium .6-.8 mm. in diameter, the stipe about the same length. Fries considered this to be a mere variety of *D. farinaceum*, but it is readily distinguished by its very small spores.

B. *Sporangia sessile*.

10. Didymium serpula, Fr. Plasmodium yellow. Plasmodiocarp much depressed, subrotund or usually more or less [Pg 64] elongated, bent, flexuous and reticulate; the wall dark-colored, with a thin layer of stellate crystals of lime. Columella entirely wanting. Capillitium of very slender threads, extending from base to upper surface, much branched, the branches combined into a dense network; to these threads adhere numerous roundish vesicles, composed of a brownish membrane, inclosing a yellow coloring matter, the vesicles 30–50 mic. in diameter. Spores globose, very minutely warted, violaceous, 7–8 mic. in diameter.

Growing on old leaves, bark, etc. The plasmodiocarp .6-.8 mm. in thickness and extending from one to several millimeters in length. This species is reported from the United States by Massee. It ought to be readily recognized by its yellow plasmodium and the peculiar vesicles adherent to the capillitium.

11. Didymium anellus, Morgan, n. sp. Plasmodiocarp in small rings or links, then confluent and elongated, irregularly connected together, bent and flexuous, resting on a thin venulose hypothallus; the wall firm, dark-colored, with a thin layer of stellate crystals of lime, irregularly ruptured. Columella merely a thin layer of brown scales. Capillitium of slender dark-colored threads, which extend from base to wall, more or less branched, and combined into a loose net. Spores globose, very minutely warted, violaceous, 8–9 mic. in diameter. Plate XII, Fig. 41.

Growing on old leaves in woods in Spring. Plasmodiocarp in rings .3-.5 mm. in diameter, or often more or less elongated into links and chains, which are bent and flexed in quite an irregular manner, the thread or vein composing them about .2 mm. in thickness. A more minute species than *Didymium serpula*, without characteristic thickenings upon the threads of the capillitium, and wanting the peculiar large cells of this species.

II. SPUMARIA, Pers. Æthalium composed of numerous elongated, irregularly-branched sporangia, more or less closely compacted together and confluent, seated upon a well-developed common hypothallus; the walls of the sporangia [Pg 65] a thin membrane with an outer layer of minute, stellate crystals of lime. Each sporangium traversed by a central subcylindric hollow columella, which extends also to the branches, but does not reach to their apices. Capillitium of slender threads, more or less branched, and combined into a network. Spores globose, violaceous.

Spumaria is essentially related to Didymium by the crystals of lime upon the walls of the sporangia. Rostafinski's figure 158 can only be regarded as ideal or diagrammatic. I am disposed to question the existence of the central columella altogether; if it does exist, it must be extremely defective.

1. Spumaria alba, Bull. Plasmodium white, amplectant. Æthalium variable in form and size, resting upon a white, membranaceous hypothallus, and usually covered by a white, friable, common cortex composed of minute crystals of lime. The component sporangia elongated, irregular, more or less branched, the branches rude, deformed, compressed, laterally confluent, obtuse or pointed at the apex; the walls of the sporangia thin and delicate, rugulose, pellucid, with a tinge of violet, iridescent when divested of the crystals of lime. Capillitium of slender threads, more or less branched and combined into a net; the threads dark colored, with pellucid extremities, and furnished with occasional rings or roundish swellings throughout their length. Spores globose, densely spinulose, dark violaceous, 10–14 mic. in diameter. Plate XII, Fig. 43.

Climbing up and surrounding the stems of small shrubs, herbaceous plants, culms of grasses, etc., especially those of living plants, rarely effused upon old wood, bark, leaves, etc. The æthalium from two or three to several centimeters in length, and with a radial thickness of two or three to several millimeters. The following forms or varieties have been distinguished as species at different times:

Var. 1. DIDYMIUM. Sporangia irregular, simple or lobed and branched, lifted up on narrow, flat extensions of the hypothallus, as if furnished with short white stipes; the common cortex wanting. This is *Didymium spumarioides*, Fr.; it is probably a dwarf form of the next variety. Plate XII, Fig. 42.

Var. 2. CORNUTA. Æthalium large and rugged in outline, cinerous from the scanty cortex; the sporangia loosely com [Pg 66] pacted, the branches running out into numerous free-pointed extremities. Capillitium of rather thick threads, forming a dense net, with broad expansions at the angles. Spores 11–14 mic. in diameter. This is *Spumaria cornuta*, Schum. It is evidently the form so elaborately figured by Rostafinski, and which Fries says abounds in Northern Europe.

Var. 3. MUCILAGO. Æthalium large, even and uniform in outline, covered by a thick, white, common cortex; the sporangia laterally confluent and densely compacted together throughout. Capillitium of rather slender threads, forming a loose net, scarcely expanded at the angles. Spores 10–13 mic. in diameter. This is *Spumaria mucilago*, Nees, as figured by Greville in the Scottish Cryptogamic Flora. The capillitium is figured by McBride in The Myxomycetes of Iowa. This is the only form I have met with in this country.

III. DIDERMA, Pers. Sporangia subglobose and stipitate or more often sessile, sometimes plasmodiocarp; the wall a thin membrane, with an outer layer composed of minute roundish granules of lime, which are usually compacted into a smooth continuous crust. Stipe present or mostly absent; the columella usually well developed. Capillitium of very slender threads, stretching from the columella to the wall of the sporangium, more or less branched, and combined into a loose net by short lateral branchlets. Spores globose, violaceous.

This genus is easily recognized by the smooth crustaceous layer of lime on the outer surface of the sporangium; in many cases this easily shells off or breaks away. Such a coating occurs in a few species of Physarum, but here the vesicles of lime attached to the threads distinguish them. This is Chondrioderma of Rostafinski's monograph; the reason for coining a new name and entirely discarding the old one is not apparent.

§1. Leangium. Sporangium usually stipitate; the wall at maturity separating from the inner mass of spores and capillitium and splitting in a stellate manner, the segments becoming reflexed. [Pg 67]

1. Diderma radiatum, Linn. Sporangium depressed-globose, the base flattened or umbilicate, stipitate or nearly sessile; the wall smooth, whitish or pale brown, splitting from the apex downward into a few reflexed irregular segments. Stipe short, thick, erect, tapering downward, standing on a thin membranaceous hypothallus; the columella large, convex, globose or obovoid, roughened. Capillitium of slender dark-colored threads, radiating from the columella, simple or branching outwardly. Spores globose, minutely warted, dark violaceous, 8-10 mic. in diameter.

Growing on old bark and wood. Sporangium .8-1.2 mm. in diameter, the stipe shorter than the diameter, sometimes nearly obsolete. Apparently rare in this country.

2. Diderma floriforme, Bull. Sporangium globose or obovoid, stipitate, growing closely crowded together on a thin brown membranaceous hypothallus; the wall smooth, varying in color from whitish or yellowish to bright brown, splitting into irregular segments, which become reflexed and revolute. Stipe long, erect, white or yellowish to brown; the columella elongated, obovoid to clavate, roughened, colored as the stipe. Capillitium of dark-colored threads, radiating from the columella and sparingly branched. Spores globose, with minute scattered warts, dark violaceous, 9-11 mic. in diameter. Plate XII, Fig. 44.

Growing on old wood of oak, hickory, etc., late in Autumn. Sporangium .7-1.0 mm. in diameter before dehiscence, the stipe usually longer than the sporangium. The color of stipe, columella and sporangium varies from white through yellowish to brown; the spores are quite peculiar by reason of their prominent scattered warts. I do not think *Chondrioderma Lyallii*, Massee, can be maintained as a separate species.

§2. Stromnidium. Sporangia growing closely crowded together upon a thick highly-developed calcareous common hypothallus, either seated upon it or partially imbedded in its substance; the wall rupturing irregularly.

3. **Diderma spumarioides, Fr.** Sporangia rather small, irregularly subglobose, sessile, seated close together on a strongly-developed whitish or yellowish common hypothallus; the wall white, rugulose, covered by a dense farinaceous layer [Pg 68] of lime. Columella convex, roughened, white or yellowish, sometimes scarcely developed. Capillitium rather scanty, of slender colorless threads, sparingly branched, ascending from the columella. Spores globose, minutely warted, violaceous, 8–10 mic. in diameter.

Growing on old leaves, bark, moss, etc. Sporangia .4–.6 mm. in diameter, irregular and rugulose. On account of the pulverulent coat of lime on the sporangium, Massee refers the species back to Didymium, where it was placed by Fries.

4. **Diderma stromateum, Link.** Sporangia large, subglobose, depressed, irregular and angular from mutual pressure, closely crowded together on a thick yellowish or pinkish common hypothallus; the wall smooth and even, grayish-white or cinereous, with a thin closely connate outer layer of minute granules of lime. Columella considerably elevated or much depressed, convex, subglobose or quite irregular, white or colored, as the hypothallus, especially at the base. Capillitium of abundant colored threads, more or less branched and combined into a loose net. Spores globose, minutely warted, violaceous, 8–10 mic. in diameter.

Growing on Hickory bark. The sporangia .5–.8 mm. in diameter, the surface smooth. Rostafinski, in his Monograph, places this species as a variety of *D. spumarioides*, but in the Appendix it is separated as a species. The sporangia are quite different from those of *D. spumarioides*, but I can see no difference in the spores.

5. **Diderma globosum, Pers.** Sporangia subglobose, more or less irregular from mutual pressure, closely crowded together on a thick, white hypothallus, seated upon it or usually sunk into its substance at the base; the wall with a white, smooth, and polished crustaceous outer layer of lime, distinct and separable from the inner membrane, easily breaking into fragments, and falling away: the inner membrane very thin, rugulose, cinereous with granules of lime or free from them and iridescent. Columella white, small, irregular, subglobose or ellipsoidal, rarely wanting. Capillitium of slender, dark colored threads, more or less branched and combined

into a loose net. Spores globose, very minutely warted, violaceous, 8-10 mic. in diameter. [Pg 69]

Growing on old leaves. Sporangia .5-.8 mm. in diameter, the surface smooth and polished.

6. Diderma crustaceum, Peck. Sporangia subglobose, irregular from mutual pressure, closely crowded together on a thick, yellowish-white common hypothallus, and at the base confluent with its substance; the wall with a creamy white, smooth, crustaceous outer layer of lime, distinct and separable from the inner membrane, and easily breaking up and falling away; the inner membrane very thin, rugulose, cinereous and iridescent. Columella whitish or cream colored, small, irregular, subglobose or ellipsoidal, often wanting. Capillitium of slender, uneven, dark colored threads, branched and combined into a loose net. Spores globose, minutely warted, violet-black, opaque, 12-15 mic. in diameter. Plate XII, Fig. 45.

Growing on old leaves, sticks, etc. A common species in this country. Sporangia .7-1.0 mm. in diameter, the surface even but finely pulverulent rather than polished. *Chondrioderma affine*, Rost., is said to be the same species. It is readily distinguished from *D. globosum*, by the much larger spores.

§3. Astrotium. Sporangia gregarious, scattered, or sometimes crowded and confluent, often much depressed, sessile, rarely stipitate; the hypothallus none or very scanty.

7. Diderma Michelii, Lib. Sporangia orbicular, very much depressed, often umbilicate above and concave underneath, stipitate or sessile, gregarious, with the margins of the sporangia sometimes confluent. The wall with a white crustaceous layer of lime, which soon ruptures around the edges, allowing the upper part to break in pieces and fall away; the inner membrane cinereous, rupturing irregularly. Stipe short, stout, erect, arising from a small, circular hypothallus, whitish or alutaceous, longitudinally rugulose, expanding at the apex, the wrinkles running out as veins on the under side of the sporangium; the columella much flattened, lenticular or discoid, alutaceous or pinkish. Capillitium of very slender, colorless threads, simple or forking a time or two, and connected by short branchlets at the extremities. Spores globose, even, violaceous, 7-9 mic. in diameter. [Pg 70]

Growing on sticks, leaves, herbaceous stems, etc. Sporangium .8–1.2 mm. in diameter, the stipe shorter than the diameter, sometimes very short or quite obsolete.

8. Diderma testaceum, Schr. Sporangia circular or oval, much depressed, sessile, without any hypothallus, gregarious, irregularly scattered, sometimes close and even confluent. The outer calcareous layer of the wall thick, smooth, crustaceous, separate and distinct from the inner membrane, white or pinkish-white to rose-red in color, gradually breaking up in pieces and falling away; the inner membrane thin, pellucid, cinereous from the adherent granules of lime, irregularly dehiscent from the apex downward. Columella hemispheric or depressed, granulose-roughened, white, pinkish, or fleshy-red. Capillitium of very slender, nearly colorless threads, more or less branched. Spores globose, very minutely warted, 8–10 mic. in diameter.

Growing on old leaves, wood, mosses, etc. Very common in this country. Sporangium .6–1.0 mm. in diameter, sometimes a little irregular, especially the form growing on mosses, and occasionally confluent.

9. Diderma cinereum, Morgan, n. sp. Sporangia subglobose, more or less irregular, somewhat depressed, sessile, usually close or crowded, sometimes confluent; the hypothallus a thin membrane, pellucid or with occasional patches of lime granules, sometimes not apparent. The wall very thin, even or rugulose, cinereous, the thin membrane covered by a single layer of closely-adherent granules of lime, rupturing irregularly. Columella white, hemispheric or depressed and irregular, the surface granulose. Capillitium of very slender, colored threads, the extremities pellucid, more or less branched. Spores globose, minutely warted, violaceous, 9–11 mic. in diameter. Plate XII, Fig. 46.

Growing on old wood, leaves, etc. The sporangium .3–.5 mm. in diameter, thin and smooth or rugulose. The species superficially greatly resembles *Physarum cinereum*.

10. Diderma difforme, Pers. Plasmodiocarp roundish, oblong, or more or less elongated and flexuous, scattered or seriately disposed; the hypothallus a thin brownish membrane, or commonly not apparent. The outer calcareous [Pg 71] layer of the wall snow-white,

thin, smooth, distinct from the inner membrane, breaking into pieces and falling away; the inner membrane thin, opaque and bluish or pellucid and iridescent. Columella reduced to a thin layer of scales and granules upon the brownish basal membrane. Capillitium scanty, consisting of short nearly colorless threads, which are simple, or fork a time or two. Spores globose, even, dark violaceous, 10–13 mic. in diameter.

Growing on bark, leaves, twigs, herbaceous stems, etc. Plasmodiocarp .3-.5 mm. in thickness and variable in length, sometimes elongated several millimeters.

11. Diderma reticulatum, Rost. Plasmodiocarp very much depressed, roundish, oblong, much elongated and flexuous, closely crowded together and confluent; the hypothallus a thin white granulose layer of lime, scarcely broader than the plasmodiocarp. The outer calcareous layer of the wall white, distinct, very fragile and easily shelling off; the inner membrane much shrunken and withdrawn from the outer coat, rugulose, cinereous, with a dense closely-adherent layer of granules of lime. The columella a thin alutaceous, granulose-roughened layer, extending along the base of the plasmodiocarp. Capillitium of threads short and very slender, colorless, somewhat branched. Spores globose, even, pale violaceous, 6–8 mic. in diameter. Plate XII, Fig. 47.

Growing on old wood, leaves, twigs, etc. Plasmodiocarp .5-.8 mm. in width, much flattened and usually closely crowded. The rough calcareous base of the plasmodiocarp might be considered as either all columella or all hypothallus, with the upper surface leather-colored. I am indebted to Arthur Lister, of London, for the determination of my specimens.

12. Diderma effusum, Schw. Plasmodiocarp very much flattened, longitudinally creeping and reticulate or altogether widely effused; hypothallus none. The wall very thin, smooth, white or cinereous, the thin membrane covered by a single layer of closely-adherent granules of lime, rupturing irregularly. The columella reduced to a thin alutaceous layer of granules of lime, forming the base of the plasmodiocarp. Capillitium of short colorless threads, extending from base to [Pg 72] wall, the extremities branched and connected

together. Spores globose, even, pale violaceous, 8–10 mic. in diameter. Plate XII, Fig. 48.

Growing on old leaves. The plasmodiocarp forms very much flattened irregular patches from a few to several millimeters in length or extent. I am indebted to Dr. Geo. A. Rex, of Philadelphia, for the identification of my specimens, with those in the herbarium of Schweinitz, under the name of *Physarum effusum*.

IV. LEPIDODERMA, DeB. Sporangium stipitate or sessile, sometimes plasmodiocarp; the wall a thin, firm, colorless membrane, with an outer layer of large scales consisting of bicarbonate of lime, the scales either lying upon the wall or inclosed in lenticular cavities of the membrane. Stipe present or absent; the columella usually conspicuous. Capillitium of very slender threads, simple or outwardly branching at a sharp angle, connected at the extremities. Spores globose, violaceous.

"In the present genus the carbonate of lime is present in the form of very minute amorphous lumps until near to maturity, when it is dissolved and reappears as bicarbonate of lime deposited in comparatively large flakes."—*Massee*.

1. Lepidoderma tigrinum, Schr. Sporangium large, much depressed, hemispheric or lenticular, the base umbilicate, stipitate; the wall a firm, dark colored membrane, variegated with large and small irregular shining scales, greenish-yellow or straw color, rupturing irregularly. Stipe stout, thick, erect, rugulose, ochraceous or ferruginous, variable in length, expanding at the base into a thin hypothallus; the columella brown, convex or hemispheric. Capillitium of slender, dark colored threads, simple or sparingly branched, radiating from the columella to the wall. Spores globose, minutely warted, dark violaceous, 10–13 mic. in diameter.

Growing on old wood, moss, etc. Sporangium 1–1.5 mm. in diameter, the stipe 1 mm. or less in length. This appears to be the only species of the genus thus far discovered in this country.

EXPLANATION OF PLATE XII.

- Fig. 37.—Didymium proximum, B. & C. *a*. Sporangium and stipe × 33. *b*. Section through the columella.

- Fig. 38. — Didymium eximium, Peck. *a.* Showing the rough columella of one form. *b.* Section through the discoid columella of the very much depressed form. Magnified by 33.
- Fig. 39. — Didymium minus, Lister. *a.* Sporangium and stipe × 33. *b. c. d.* Sections through the columella showing different forms.
- Fig. 40. — Didymium farinaceum, Schr. Section through the columella. After Rostafinski.
- Fig. 41. — Didymium anellus, Morgan, *a.* Growing upon a leaf × 3. *b.* Plasmodiocarp × 17.
- Fig. 42. — Spumaria alba, Bull. Var. 1. didymium, sporangia × 3. Drawn from a foreign specimen.
- Fig. 43. — Spumaria alba, Bull. *a.* Æthalium natural size. *b.* Capillitium and spores as seen by a magnifying power of 500 diameters.
- Fig. 44. — Diderma floriforme, Bull. Stipe and columella × 20.
- Fig. 45. — Diderma crustaceum, Peck. *a.* Sporangia crowded on the thick hypothallus, natural size. *b.* Sporangia × 11. *c.* Section through outer coat, inner membrane, and columella.
- Fig. 46. — Diderma cinereum, Morgan, *a.* Sporangia growing on a leaf × 3. *b.* Sporangia × 23. *c.* Section through the wall and columella.
- Fig. 47. — Diderma reticulatum, Rost. Plasmodiocarp growing on leaf × 3.
- Fig. 48. — Diderma effusum, Schw. Plasmodiocarp effused on a leaf × 3.

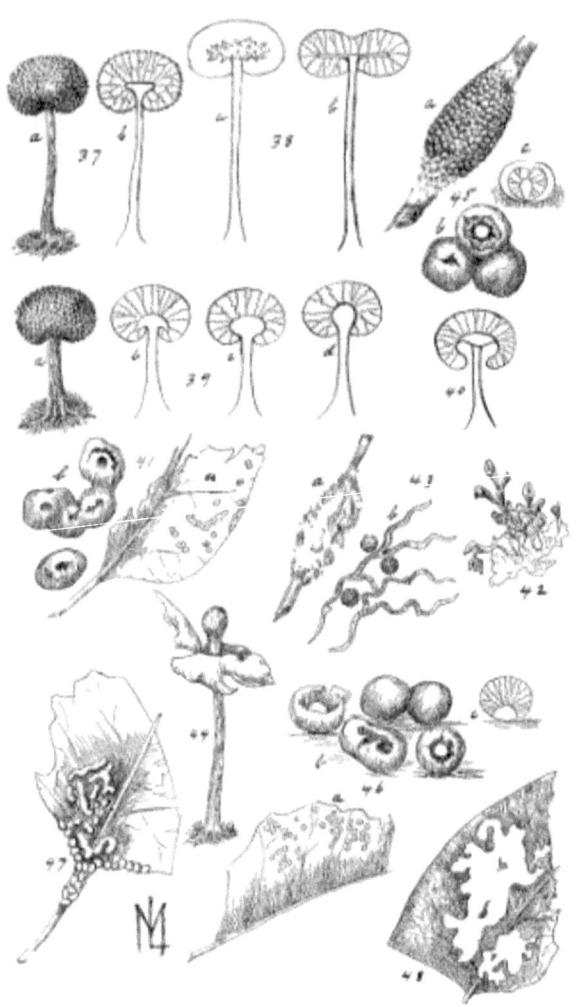

The Journal of the Cin. Soc. Natural History
Vol. XVI. Plate XII.

[Pg 73]

Reprint from The Journal of the Cincinnati Society of Natural History, August, 1896.

THE MYXOMYCETES OF THE MIAMI VALLEY, OHIO.

By A. P. Morgan.
Fourth Paper.
(Read May 6, 1896.)

Order VIII. PHYSARACEÆ.

Sporangia simple and stipitate or sessile, sometimes plasmodiocarp, rarely combined into an aethalium; the wall a thin membrane, usually with an outer layer of minute roundish granules of lime. Stipe present or often wanting, seldom prolonged within the sporangium as a columella. Capillitium consisting of slender tubules, which branch repeatedly in every direction and anastomose to form an intricate network, the extremities attached on all sides to the wall of the sporangium; the tubules more or less expanded at the angles of the network and inclosing minute roundish granules of lime, these granules either aggregated into nodules with intervening empty spaces or more rarely distributed throughout their entire length. Spores globose, very rarely ellipsoidal, violaceous.

This order is at once distinguished from the Didymiaceæ by the presence of the granules of lime in the capillitium. [Pg 74]

Table of Genera of Physaraceæ.

I. Tubules of the capillitium having the granules of lime in them aggregated into roundish or angular nodules, with intervening empty spaces.

A. Outer surface of the sporangium destitute of lime.

- 1. Angioridium. Plasmodiocarp laterally compressed, splitting regularly into two valves.
- 2. Cienkowskia. Plasmodiocarp terete, elongated, irregularly dehiscent.
- 3. Leocarpus. Sporangia subglobose or obovoid, stipitate or sessile.

B. *Outer surface of the sporangium invested with granules of lime.*

   a. *Stipe prolonged within the sporangium as a columella.*

- 4. Physarella. Sporangium oblong, stipitate, the apex re-entrant.
- 5. Cytidium. Sporangium globose, stipitate, the apex convex.

   b. *Stipe never entering the sporangium.*

- 6. Craterium. Sporangium obovoid to cylindric, stipitate.
- 7. Physarum. Sporangium globose, depressed globose or irregular, stipitate or sessile.
- 8. Fuligo. Aethalium a compound plasmodiocarp.

II. Tubules of the capillitium with the granules of lime in them distributed throughout their entire length.

- 9. Badhamia. Stipe not prolonged within the sporangium as a columella.
- 10. Scyphium. Stipe entering the sporangium and prolonged within it as a columella. [Pg 75]

I. ANGIORIDIUM Grev. Plasmodiocarp laterally compressed, more or less elongated and flexuous, attached by the lower margin to the substratum, and, at maturity, regularly dehiscent along the upper margin by a longitudinal fissure; the wall a firm membrane, with the granules of lime forming a reticulate layer on the inner surface. Capillitium a loose, irregular net-work of tubules, extending from side to side, and containing large, irregular nodules of lime. Spores globose, violaceous.

A genus readily distinguished by its laterally compressed plasmodiocarp, splitting lengthwise by a regular fissure. The wall is a single membrane, and there is but a single reticulate layer of lime upon it, which is plainly on the inner surface.

1. Angioridium sinuosum Bull. Plasmodiocarp laterally compressed and very much flattened, more or less elongated and flexuous, sometimes confluent and branched or reticulate, without any hypothallus; the wall a more or less thickened and brownish membrane, the inner surface coated with a dense reticulately thickened white layer of lime, and often studded with the white nodules. Capillitium of hyaline tubules, forming a loose irregular net-work, with numerous broad vesicular expansions filled with lime; the nodules white, very large, irregularly lobed, and branched. Spores globose, very minutely warted, violaceous, 8–10 mic. in diameter.

Growing on old leaves, sticks, mosses, etc. Plasmodiocarp commonly about 1 mm. in height and 1–5 mm. in length, but the size is variable. The color appears to depend upon the thickening of the membrane; when it is thin and pellucid, the color is white or cinereous from the inner layer of lime and the contained spores; with a more thickened membrane, the color becomes ochraceous or brownish. *Physarum bivalve* Pers. *Physarum sinuosum* of Rostafinski's monograph. See Plate XIII. Fig. 49.

II. CIENKOWSKIA Rost. Plasmodiocarp terete, elongated, flexuous, creeping, and reticulate, irregularly dehiscent; the wall a more or less thickened membrane, externally naked, with the granules of lime on the inner surface. Capillitium [Pg 76] of slender tubules, combined into an irregular network, attached on all sides to the wall of the sporangium, and bearing everywhere short pointed or uncinate free branchlets; the lime in thin transverse plates and irregular nodules. Spores globose, violaceous.

The peculiar characteristic of this genus is the short free hooked and pointed branchlets of the capillitium.

1. Cienkowskia reticulata A. & S. Plasmodiocarp more or less elongated, curved and flexuous, simple or branched, sometimes confluent and reticulate, breaking away first along the upper surface, leaving an irregular margin. The wall a firm yellow membrane, with thinner hyaline areas and with thicker yellow-brown or red-brown spots; the outer surface without any lime, smooth, and shining; the inner surface with a dense layer of yellow granules raised at intervals into transverse ridges, these are connected with broad thin flat plates of lime which traverse the capillitium, forming imperfect

septa to the sporangium. Capillitium consisting of slender yellow tubules, forming a network of irregular meshes, with slight expansions at the angles and bearing along the sides short pointed or uncinate free branchlets; the tubules containing a few scattered yellow nodules of lime various in size and shape. Spores globose, very minutely warted, violaceous, 8–10 mic. in diameter.

Growing on old wood, bark, leaves, etc. Plasmodiocarp in veins .3-.5 mm. in thickness, sometimes forming a net-work a centimeter or more in extent. This curious Myxomyces seems very rare in America. I have met with it but once. The specimen in the herbarium of Schweinitz, marked *Physarum reticulatum*, is not this species, though it answers well enough to the original description. See Plate XIII. Fig. 50.

III. LEOCARPUS Link. Sporangia subglobose or obovoid, stipitate or sessile; the wall a more or less thickened membrane, the external surface destitute of lime, polished and shining, irregularly dehiscent. Stipe short, poorly developed or sometimes wanting. Capillitium of slender tubules, forming an irregular net-work more or less expanded at the angles; [Pg 77] the tubules enlarging at intervals into vesicles, which usually contain nodules of lime. Spores globose, violaceous.

A genus characterized by the form of the sporangia and the smooth and glossy surface of the wall.

1. Leocarpus psittacinus Ditm. Sporangium small globose or somewhat depressed, stipitate or subsessile; the wall a thin membrane, rugulose and iridescent, with thicker red or yellow spots and patches, destitute of lime. Stipe weak, erect or inclined, variable in length, the base expanded, orange to red in color. Capillitium a dense net-work of tubules, widely expanded at the angles and bearing numerous irregular vesicles, various in size and form, yellow or orange to red in color. Spores globose, even, dark violaceous, 7–9 mic. in diameter.

Growing on old wood, leaves, etc. The sporangium .5-.6 mm. in diameter, the stipe about the same length or sometimes very short. The sporangia are dull brownish to the naked eye, but when magnified the green, purple, and blue metallic tints of the wall become apparent. There does not appear to be any granules of lime either on the wall or in the capillitium. *Physarum psittacinum* Ditm.

2. Leocarpus cæspitosus Schw. Sporangium small subglobose or obovoid to turbinate, somewhat irregular, stipitate or subsessile; the wall a reticulately thickened and fragile membrane, yellow-brown to greenish-yellow or olivaceous in color, externally rugulose and glossy, the inner surface with scales and patches of lime. Stipe short and thick, sometimes nearly obsolete, yellowish or reddish brown, darker below, the base expanded into a small hypothallus. Capillitium a loose irregular net-work of tubules with wide expansions at the angles; the nodules of lime large, numerous, white or yellowish, irregular, with acute angles and pointed lobes. Spores globose, minutely warted, dark violaceous, 9–11 mic. in diameter.

Growing cæspitosely or scattered on old wood and mosses. Sporangium .6-.8 mm. in diameter, variable in shape, the stipe usually very short. *Physarum cæspitosum* Schw., *North American Fungi*. My specimens, some of them, have been referred to *Physarum citrinellum* Peck; others to *Physarum variabile* Rex. See Plate XIII. Fig. 52. [Pg 78]

3. Leocarpus brunneolus Phillips. Sporangium large, globose or somewhat depressed, sessile; the wall a thick yellow-brown membrane, the outer surface naked, smooth, and polished, with a dense white inner layer of granules of lime, dehiscing in a stellate manner, the segments becoming reflexed. Capillitium of tubules forming a dense net-work, with wide expansions at the angles; the nodules of lime very large, numerous, white, angular and irregular. Spores globose, minutely warted, dark violaceous, 8–10 mic. in diameter.

Growing on bark of oak, California (*Harkness.*) Sporangium nearly 1 mm. in diameter. *Diderma brunneolum* Phillips. I have taken the description from Massee's monograph.

4. Leocarpus fragilis Dicks. Sporangium very large, obovoid-oblong, stipitate or subsessile; the wall a greatly thickened membrane, polished and shining within and without, from alutaceous or pale umber to dark-brown in color, destitute of lime. Stipe short, weak, and slender, arising from a thin hypothallus. Capillitium of slender tubules forming a loose network of large irregular meshes, with slight expansions at the angles; the lime white, variable in amount, sometimes quite scanty, then again filling large portions of

the net-work with long-branched and reticulate masses. Spores subglobose, dark violaceous, opaque, 12–15 mic. in diameter.

Growing gregariously on old wood, leaves, mosses, etc. Sporangium 1.5–2 mm. in length by 1 mm. in thickness, the stipe variable in length, but usually much shorter than the sporangium. *Diderma vernicosum* Pers. See Plate XIII. Fig. 51.

IV. PHYSARELLA Peck. Sporangium oblong, stipitate; the apex re-entrant and confluent with the hollow columella; the wall a thin membrane covered with small scales and minute granules of lime, at maturity torn away at the apex and stellately splitting into a few segments. Stipe elongated, tapering upward, entering the sporangium and prolonged to the apex as a tubaeform columella. Capillitium distinguished by two distinct sets of tubules; the first consisting of long, thick tubules filled with lime, rising at regular intervals from [Pg 79] the wall of the sporangium and extending to the columella; the second, of very slender threads, scarcely branched, and nearly destitute of lime, stretching between the wall and the columella. Spores globose, violaceous.

A genus founded upon the one remarkable species, and more distinct than any other from the typical genus of the Physaraceae. In fact, the structure of the sporangium is unique among the Myxomycetes.

1. Physarella oblonga B. & C. Sporangium oblong, the apex re-entrant and confluent with the summit of the columella, the base obtuse or slightly umbilicate, stipitate, cernuous. The wall of the sporangium a firm, yellowish membrane, covered with minute granules and with scattered, small, yellow scales of lime; after maturity the apex is torn away more or less irregularly from the summit of the columella and the wall splits into a few segments, which become reflexed and are subpersistent about the base of the sporangium. Stipe long, erect or flexuous, the apex bent or curved, red-brown, rising from a small hypothallus, entering the sporangium and prolonged to the apex as a hollow tubaeform columella. Capillitium of thick, spiniform tubules filled with lime and slender, violet threads, extending between the wall and the columella. The tubules elongated, terete, tapering gradually from wall to columella, containing yellow granules of lime; the threads very slender, outwardly

branched a time or two, the further extremities connected by short, lateral branches, often furnished with minute, free branchlets, and containing a few small, fusiform nodules of lime. Spores globose, nearly smooth, violaceous, 7-9 mic. in diameter.

Growing on old wood, bark, leaves, etc. Sporangium commonly .8-1.0 mm. in length by .5-.6 mm. in diameter, the stipe 1-2 mm. long; the spiniform tubules measure 150-200 × 15-20 mic. See Plate XIII. Fig. 53.

The abnormal forms of this species which sometimes manifest themselves are very singular; the sporangium has a tendency to dilate, becoming funnel-form or even salver-shaped, the stipe shortening and even disappearing. I have a large specimen which superficially resembles some lichen, a *Physcia*, for example; the sporangia are pressed down, flattened out, extremely irregular, and in many places confluent; [Pg 80] the rudimentary stipes are hidden beneath the leafy expansions. In all the forms, however, may be uncovered the spiniform tubules mingled with the slender threads. This is *Trichamphora oblonga* B. & C. *Tilmadoche oblonga* of Rostafinski's monograph, and *Physarella mirabilis* Peck.

V. CYTIDIUM Morgan. Gen. nov. Sporangium globose or rarely ellipsoidal, stipitate; the wall a thin membrane, with an external layer of minute granules of lime, rupturing irregularly. Stipe more or less elongated, tapering upward and entering the sporangium as a columella. Capillitium of slender tubules, arising from the columella, repeatedly branching and anastomosing to form a regular net-work, the extremities attached on all sides to the wall of the sporangium, the tubules containing at intervals nodules of lime. Spores globose, violaceous.

This genus is readily distinguished from *Physarum* by the columella, which gives origin to the capillitium; this feature indicates a relationship to *Didymium* and to *Lamproderma*.

§1. Eucytis. Sporangium globose, the columella not reaching its center.

1. Cytidium pulcherrimum B. & R. Sporangium globose, stipitate; the wall a thin lilac-tinted membrane, with a dense closely adherent layer of granules of lime, dark purple or wine-colored. Stipe long,

erect, dark purple to purplish black, tapering upward and entering the sporangium as a slight obtuse columella. Capillitium of slender lilac tinted threads, forming a dense net-work of very small meshes, with slight expansion at the angles; the nodules of lime very small, numerous, dark purplish or vinose in color, ellipsoidal or obtusely angular. Spores globose, even, lilac, 7–9 mic. in diameter.

Growing on old wood. Sporangium .4-.5 mm. in diameter, the stipe two or three times as long; the lime-nodules about the size of the spores. The purple stain, which the sporangia leave on white paper, is made by the granules of lime; the [Pg 81] spores color the paper violet. *Physarum pulcherrimum* B. & Rav., and *P. atrorubrum* Peck.

2. Cytidium citrinum Schum. Sporangium globose, the base slightly flattened or umbilicate, stipitate; the wall a thin membrane, covered with small scales of lime, yellow or greenish-yellow, breaking up and falling away at maturity. Stipe stout, erect, yellow, longitudinally rugulose, expanded at the base, tapering upward and entering the sporangium as a short obtusely conical columella. Capillitium of slender tubules, forming a dense net-work, with slight expansions at the angles; the lime-nodules numerous, roundish or ellipsoidal, variable in size, yellow. Spores globose, nearly smooth, violaceous, 7–8 mic. in diameter.

Growing on bark, leaves, mosses, etc. Sporangium .5-.6 mm. in diameter, the stipe from once to twice this length. This, the typical species, I have not seen in this country, but forms with the sporangium lemon-yellow and grayish-yellow, with the stipe golden-yellow, connect it with *C. rufipes*. It is *Physarum citrinum* Schum. *Diderma citrinum* of Fries., S. M.

3. Cytidium rufipes A. & S. Sporangium globose, sometimes a little depressed and the base umbilicate, stipitate; the wall a thin membrane, covered with small scales of lime, golden-yellow to orange in color, breaking up at maturity and falling away. Stipe variable in length, slender, from orange or orange-red to dark red in color, sometimes blackish below, rising from a thin hypothallus, tapering upward and entering the sporangium as a short obtuse columella. Capillitium of slender tubules, forming a dense net-work of very small meshes, slightly expanded at the angles; the nodules

of lime small, numerous, ellipsoidal or obtusely angular, orange to red in color. Spores globose, nearly smooth, violaceous, 8–10 mic. in diameter.

Growing on old wood, mosses, etc. A very abundant species. Sporangium .5–.7 mm. in diameter, the stipe from once to twice as long. As here defined, the species includes *Physarum aurantium* var. *rufipes* A. & S., and *Physarum aureum* var. *chrysopus* Lev, which I am unable to keep separate; the variation in size of the spores is not in correspondence with the variations in color of the sporangia. *Physarum [Pg 82] pulchripes* Peck, and *Physarum petersii* B. & C., mostly belong here. The bright orange colors become dull or tawny with age and exposure to the weather.

4. Cytidium ravenelii B. & C. Sporangium globose, stipitate; the wall a thin pellucid membrane, covered with small scales of lime, from gray or drab to pale umber in color, breaking up at maturity and falling away. Stipe variable in length, concolorous with the sporangium or darker below, tapering upward and entering the sporangium as a short obtusely conical columella. Capillitium of tubules, forming a dense net-work of very small meshes, with slight expansions at the angles; the lime-nodules small, numerous, ellipsoidal or obtusely angular, gray or drab to pale umber in color. Spores globose, nearly even, pale violaceous, 7–9 mic. in diameter.

Growing on old wood, mosses, etc. Sporangium about .5 mm. in diameter, the stipe once to twice this length. The species as here described includes *Didymium ravenelii* B. & C., *Physarum simile* Rost., and *Physarum murinum* Lister.

5. Cytidium globuliferum Bull. Sporangium globose, the base sometimes flattened or slightly umbilicate, stipitate; the wall a thin, pellucid membrane, covered with small scales of lime, white, cream-colored, or sometimes pinkish, breaking up and falling away at maturity. Stipe variable in length, white or smoky-white, usually darker below, rising from a thin hypothallus, tapering upward and entering the sporangium as a short obtuse or conical columella. Capillitium of slender tubules, forming a dense, persistent net-work of very small meshes, more or less expanded at the angles; the nodules of lime variable in size, numerous, white, roundish, ellipsoidal

or obtusely angular. Spores globose, nearly even, pale violaceous, 7–9 mic. in diameter.

Growing on old wood, bark, mosses, etc. A very common and abundant species. Sporangium .5-.6 mm. in diameter, the stipe from once to two or three times this length. The lime nodules in the capillitium are sometimes round and quite minute, then again they are large and obtusely angular; the columella varies from very short and conical to longer and more cylindric. *Diderma globuliferum* of Fries S. M., [Pg 83] *Physarum albicans* Peck. The specimens with the columella well nigh obsolete, may be *Tilmadoche columbina* Rost. See Plate XIII. Fig. 55.

6. Cytidium melleum B. & Br. Sporangium globose, stipitate or subsessile; the wall a thin yellowish membrane, rugulose, covered by large irregular scales of lime, honey-color to golden-yellow, breaking up irregularly. Stipe short, sometimes very short or nearly obsolete, snow-white, expanding at the base into a small white hypothallus, tapering upward and entering the sporangium as a short obtusely conical columella. Capillitium a loose net-work of delicate tubules with broad vesicular expansions containing much lime; the nodules numerous, white or sometimes yellow, large, irregular, lobed, and branched. Spores globose, nearly even, pale violaceous, 7–9 mic. in diameter.

Growing on old leaves, sticks, herbaceous stems, etc.; not uncommon in this region. Sporangium .4-.5 mm. in diameter, the stipe about the same length or much shorter. *Didymium melleum* B. & Br. *Didymium chrysopeplum* B. & C. also belongs here and not with *C. citrinum*.

§2. Rexiella. Sporangium ellipsoidal or pyriform, the columella prolonged nearly to the apex of the sporangium.

7. Cytidium penetrale Rex. Sporangium ellipsoidal or pyriform, stipitate; the wall a thin pellucid membrane, covered with small scales of lime, yellow-gray to greenish-yellow, rupturing at maturity into two to four segments. Stipe long, slender, translucent, pale red to dark red in color, tapering upward, entering the sporangium and prolonged nearly to the apex as a slender columella. Capillitium of very slender tubules, radiating from numerous points of the columella, forming a delicate net-work of very small meshes, scarce-

ly expanded at the angles; the nodules of lime small, not numerous, roundish or obtusely angled, white or yellowish. Spores globose, very minutely warted, pale violaceous, 5.5–6.5 mic. in diameter.

Growing on old wood. A rare and singular species. Sporangium .5-.7 mm. in height by .3-.5 mm. in diameter, the stipe two or three times the height of the sporangium. There is an affinity between this species and the *Physarella*. The obscure *Tilmadoche hians* Rost., may be the same as the present species. See Plate XIII. Fig. 54.

EXPLANATION OF PLATE XIII.

- Fig. 49. — Angioridium sinuosum Bull. *a*. Plasmodiocarp × 5 *b*. Capillitium and spores × 500.
- Fig. 50. — Cienkowskia reticulata A. & S. *a*. Plasmodiocarp × 5. *b*. Piece of plasmodiocarp × 90. *c*. Capillitium and spores × 500.
- Fig. 51. Leocarpus fragilis Dicks, *a*. Sporangia × 5. *b*. Capillitium and spores × 500.
- Fig. 52. — Leocarpus caespitosus Schw. *a*. Sporangia × 5. *b*. Capillitium and spores × 500.
- Fig. 53. — Physarella oblonga B. & C. *a*. Sporangia × 5. *b*. Sporangia × 90. *c*. Capillitium and spores × 500.
- Fig 54 — Cytidium penetrale Rex. *a*. Sporangia × 5 *b*. Sporangia and columella × 90. *c*. Capillitium and spores × 500.
- Fig. 55. — Cytidium globuliferum Bull. *a*. Sporangia × 5. *b*. Sporangia × 90. *c*. Columella × 90 *d*. Capillitium and spores × 500.

The Journal of the Cin. Soc. Natural History
Vol. XIX. Plate XIII.
Morgan on Myxomycetes

[Pg 84] VI. CRATERIUM Trent. Sporangium obovoid to cylindric, stipitate; the upper and usually greater part of the wall covered with granules of lime, the basal portion naked and more persistent. Stipe short or sometimes elongated, arising from a small circular hypothallus, longitudinally plicate, confluent above and similarly colored with the base of the sporangium. Capillitium of tubules, forming a loose network, bearing numerous large angular and irregular nodules of lime, which are often confluent along the axis of the sporangium into a pseudo-columella. Spores globose, minutely warted, violaceous.

In this genus the sporangium is commonly obovoid, with a naked base which is confluent with the stipe and similarly colored; after dehiscence there is left behind the more persistent cyathiform portion standing on the substratum.

§1. Eu-Craterium. Sporangium at maturity dehiscent in a regular circumscissile manner, the apex falling away as a lid, leaving behind the more persistent cup-shaped portion.

1. Craterium minutum Leers. Sporangium cyathiform, stipitate; the lid slightly convex, discrete from the first, usually depressed below the rim of the cup, falling away at maturity, and leaving a smooth, circular margin to the lower cyathiform portion. The wall a thick, firm, yellow-brown membrane, the outer surface of the cup entirely naked, smooth and shining, varying greatly in color from alutaceous or ochraceous to various shades of brown; the lid usually whitened by a thin layer of granules of lime. Stipe short, erect or bent, and slightly curved at the apex, varying in color from rusty yellow to reddish brown, longitudinally plicate, arising from a small, circular hypothallus. Capillitium of tubules forming a loose net-work, bearing large, irregular, white nodules of lime, which are sometimes confluent in the axis of the sporangium. Spores globose, very minutely warted, violaceous, 8–10 mic. in diameter.

Growing on old wood, sticks, leaves, etc. Sporangium, together with the stipe, .8–1.4 mm. in height and .3–.5 mm. in diameter, the stipe usually shorter than the sporangium, sometimes equal to it in length, rarely longer. The latest authorities include the three species *Craterium vulgare*, *C. [Pg 85] pyriforme*, and *C. minutum* of Rostafinski's monograph all in one species.

2. Craterium concinnum Rex. Sporangium usually minute, broadly funnel-shaped, stipitate; operculum always more or less convex, rarely approaching a hemispherical shape, dehiscent in a regular circumscissile manner. The wall a thick, brownish membrane, ex-

ternally smooth and variously colored, sometimes uniformly light or dark umber, sometimes dark brown below and brownish white above; the operculum brownish white, darkest in the center. Stipe short, dark brown, longitudinally ridged. Capillitium of tubules forming a close-meshed net-work, bearing small rounded or slightly angular nodules of lime, ochre-brown in color. Spores globose, very minutely warted, brown, 9–10 mic. in diameter.

Growing usually upon chestnut-burs, and frequently associated with *Lachnobulus globosus*. Sporangium .5-.8 mm. in height including the stipe and .2-.5 mm. in diameter at the top, the stipe equaling the sporangium in length. It is readily distinguished by its small nodules in the capillitium, which are invariably of a dull, brownish-ochre color.

3. Craterium rubescens Rex. Sporangium subcylindric or elongated cyathiform, stipitate; the apex convex, at maturity separating by an irregular line in a circumscissile manner. The wall dark violet-red, smooth, except at the upper portion, which is slightly roughened by an external deposit of scattered lime-granules of a pale, lilac color. Stipe short, violet-black, wrinkled longitudinally. Capillitium of tubules forming a loose, irregular net-work, bearing large, violet-red nodules of lime which are often confluent in the axis of the sporangium. Spores globose, minutely warted, dark violaceous, 7–9 mic. in diameter.

Growing on old wood, leaves, etc. Sporangium .6-.8 mm. in height including the stipe and .5-.6 mm. in diameter, the stipe one-half the height of the sporangium. The species is distinguished by the color, which exhibits some shade of red or violet-red in every part of its structure.

4. Craterium minimum B. & C. Sporangium cylindric or turbinate cylindric, stipitate; the apex convex, separating in a regular circumscissile manner by a lid. The wall a thick, [Pg 86] yellow-brown membrane, most of the outer surface covered with minute, white granules of lime, the basal portion naked. Stipe very short, plicate, red-brown, arising from a small hypothallus. Capillitium of tubules forming a loose net-work bearing large, irregular, white nodules of lime, sometimes confluent in the axis of the sporangium. Spores globose, very minutely warted, violaceous, 7–9 mic. in diameter.

Growing on old leaves, herbaceous stems, etc. Sporangium together with the stipe 1–1.5 mm. in height and .25–.35 mm. in thickness, the stipe .2–.4 mm. in length. This is a common species everywhere in the United States, and perfectly distinct from *Craterium convivale*. It is *Craterium cylindricum* of Massee's monograph, according to Lister. See Plate XIV. Fig. 56.

§2. Cupularia, Link. Sporangium irregularly dehiscent, breaking up and gradually falling away from the apex downward.

a. Stipe shorter than the sporangium.

5. Craterium convivale Batsch. Sporangium obovoid or oblong-obovoid, stipitate; the wall hyaline, thin and fragile above, the lower portion a thickened and brownish membrane, the surface, usually most of it, covered with minute white granules of lime, the base naked and brown. Stipe very short, erect, red-brown, plicate, arising from a small hypothallus. Capillitium of tubules forming a dense net-work, bearing numerous large irregular white nodules of lime, which are often confluent in the axis of the sporangium. Spores globose, very minutely warted, violaceous, 8–10 mic. in diameter.

Growing on old leaves, herbaceous stems, etc. Sporangium .6–1.0 mm. in height including the stipe and .3–.5 mm. in diameter, the stipe much shorter than the sporangium. The thin apex breaks up into pieces and falls away, leaving sometimes a regular cyathiform portion, at other times the margin is broken and irregular. This is *Craterium leucocephalum* of Rostafinski's monograph. The specimens of *Physarum scyphoides* C. & B. which I have seen appear to be a small form of this species. [Pg 87]

6. Craterium aureum Schum. Sporangium obovoid to oblong obovoid, stipitate, the wall a thin and delicate membrane above, thicker and firmer below, hyaline or yellowish, almost entirely covered by a dense layer of granules of lime, varying from lemon-yellow to orange in color. Stipe short, erect, yellow to orange, brownish toward the base, longitudinally plicate, rising from a small hypothallus. Capillitium of slender tubules, forming a dense net-work, bearing numerous rather small irregular nodules of lime, yellow or sometimes white in color, and often confluent along the axis of the sporangium. Spores globose, very minutely warted, dark violaceous, 8–10 mic. in diameter.

Growing on old leaves, sticks, herbaceous stems, etc. Sporangium and stipe .7–1.0 mm. in height and .3–.5 mm. in diameter, the stipe .2–.4 mm. long. The elongated form is the common one in this region. *Craterium mutabile* Fr.

b. Stipe longer than the sporangium.

7. Craterium nodulosum C. & B. Sporangium globose or obovoid, stipitate; the greater part of the wall a thin hyaline membrane, easily breaking away, covered externally with large white scales and nodules of lime; the basal portion naked, thickened, and more persistent, red-brown and plicate. Stipe long, erect or inclined, plicate, red-brown, rising from a small hypothallus. Capillitium of tubules forming a loose net-work, containing a variable quantity of lime in the shape of long irregular white nodules, sometimes confluent, with pointed lobes and branchlets. Spores globose, very minutely warted, dark violaceous, 10–12 mic. in diameter.

Growing on old wood, bark, leaves, etc. Sporangium .5–.6 mm. in diameter, the stipe two or three times as long. It is *Badhamia nodulosa* C. & B., *Journal of Mycology*, Vol. V, p. 186. Ravenel's specimens are on *Acacia* bark. Mr. Webber sent me elegant specimens from Florida where, he says, it grows commonly on the leaves and bark of the orange trees.

8. Craterium maydis Morgan, n. sp. Sporangium globose or obovoid, stipitate; the upper part of the wall a yellowish membrane, thin and fragile, covered with large thick scales and nodules of lime, amber-colored to golden- [Pg 88] yellow; the basal portion thicker and more persistent, naked and plicate, red-brown. Stipe red-brown, long, slender, plicate, rising from a small hypothallus. Capillitium of thick tubules, forming a net-work with wide expansions at the angles; the nodules of lime large, numerous, yellow, angularly lobed and branched. Spores globose, very minutely warted, pale violaceous, 9–10 mic. in diameter.

Growing on old stalks of *Zea mays*. Sporangium with the stipe 1–1.5 mm. in height and .4–.6 mm. in diameter, the stipe always longer than the sporangium. I find it in abundance on old stalks of Indian corn, but never on anything else. See Plate XIV. Fig. 57.

VII. PHYSARUM Pers. Sporangium globose, depressed globose or irregular, stipitate or sessile; the wall a thin membrane, with an outer layer of minute roundish granules of lime, irregularly dehiscent. Stipe present or often wanting, never prolonged within the sporangium as a columella. Capillitium of slender tubules, forming an intricate net-work, the extremities attached on all sides to the wall of the sporangium; the tubules more or less expanded at the angles of the net-work, and containing at varying intervals nodules of lime. Spores globose, violaceous.

Physarum is the central genus of the *Physaraceæ* from which all the others are detached by characters which for the most part are unimportant.

§1. Lapidium. Lime in the Capillitium scanty; the nodules small, roundish, ellipsoidal or fusiform.

A. Sporangium stipitate.

a. Sporangia regular.

1. Physarum nutans Pers. Sporangium orbicular, very much depressed, the base concave or umbilicate, stipitate, cernuous; the wall a thin pellucid membrane, thickly covered with minute white or yellow roundish scales of lime, breaking up into irregular fragments, which often remain attached to the capillitium. Stipe long, slender, tapering upward, [Pg 89] bent or curved at the apex, longitudinally rugulose, brown or blackish at the base, becoming paler upward and cinereous or whitish at the apex. Capillitium of very slender threads, rising from the base of the sporangium, forming a net-work with much elongated meshes, scarcely expanded at the angles; the nodules of lime white or yellow, ellipsoidal or fusiform, often very small and few in number, sometimes rather large and numerous. Spores globose, very minutely warted, violaceous, 8–10 mic. in diameter.

Growing on wood, bark, mosses, etc. A very common species. Sporangium .4-.5 mm. in diameter, the stipe 1–2 mm. in length, the lime-nodules commonly not thicker than the spores, but sometimes from once to twice their diameter. Under this name I have included all the lenticular species of Persoon's Synopsis, *Physarum nutans, P. luteum, P. viride* and *P. aureum*. There is no difference in these species, except in the color of the granules of lime; the form of the sporangium and the shape and color of the stipe are the same in all of them. No two authorities agree in the presentation of this species.

2. Physarum cupripes B. & R. Sporangium orbicular, much depressed, the base umbilicate, stipitate, cernuous; the greater part of the wall thin and delicate, with a scanty covering of yellow granules of lime, becoming naked and then brassy and iridescent, after maturity soon disappearing; the lower basal portion thicker and more persistent, with a layer of small yellow scales of lime. Stipe long, flexuous, bent at the apex, plicate, pale brown to yellow-brown, darker toward the base. Capillitium of slender tubules, forming a dense persistent net-work, more or less expanded at the angles; the lime-nodules small, numerous, yellow, angular and fusiform, below often confluent. Spores globose, very minutely warted, violaceous, 8–10 mic. in diameter.

Growing on old wood; rare. Sporangium .4-.5 mm. in diameter, the stipe two or three times this length. The lime nodules are found both on the sides and at the angles of the meshes, and are fusiform or angular accordingly; the lime is scanty above, but in the lower part of the capillitium the nodules sometimes run together into lobed and branched forms. This is *Physarum berkeleyi* of Rostafinski's monograph. [Pg 90]

3. Physarum obrusseum, B. &. C. Sporangium globose, the base usually slightly flattened or umbilicate, stipitate and cernuous; the wall a thin, violaceous membrane, covered by small, roundish, white or yellow scales of lime, or sometimes naked, splitting irregularly from the apex downward. Stipe long, slender, tapering upward, flexuous, bent or curved at the apex, yellow, yellow-brown, or pale brown. Capillitium of very slender tubules, forming a loose net-work, scarcely expanded at the angles; the nodules of lime small, white or yellow, roundish or obtusely angular, few to numerous, rarely wanting. Spores globose, very minutely warted, violaceous, 8–10 mic. in diameter.

Growing on old wood, bark, mosses, etc Sporangium .2-.4 mm. in diameter, the stipe 1–2 mm. in length, the lime nodules when abundant once to twice the diameter of the spores, when scanty very small. This, as I find it growing, is an extremely variable species; I think its various forms and appearances cover such species as *Didymium obrusseum* B. & C.; *D. tenerrimum* B. & C.; *Physarum tenerum* Rex, etc., etc. See Plate XIV. Fig. 58.

4. Physarum nucleatum Rex. Sporangium globose, stipitate, erect or slightly nodding; the wall a thin, pellucid membrane, thickly covered with minute, white, roundish scales of lime, which are exceptionally sparse or absent, rupturing irregularly. Stipe long, slender, yellowish-white, longitudinally rugulose, tapering upward, expanded at the base into a small hypothallus. Capillitium of very slender tubules, forming a delicate net-work of small meshes, scarcely expanded at the angles; nodules of lime small, not numerous, roundish, white, usually concentrated into a large lump in the center of the sporangium. Spores globose, very minutely warted, violaceous, 6–7 mic. in diameter.

Growing on old wood, bark, etc.; rare. Sporangium .4–.5 mm. in diameter, the stipe two or three times as long, the lime-nodules about the size of the spores. The species much resembles some of the forms of *P. obrusseum*, but is to be distinguished by its central mass of lime and the small spores.

5. Physarum compactum Wingate. Sporangium depressed-globose, the base slightly umbilicate, stipitate, cernu [Pg 91] ous; the wall a thin, violaceous membrane, rugulose and iridescent, studded with large and thick, snow-white, roundish or elliptic scales of lime, at maturity splitting from the apex downward into several segments. Stipe long, rather weak, bent and flexuous, tapering upward, longitudinally rugulose, from snow-white to whitish-ochre and smoky-white, usually brownish at the base, and arising from a thin hypothallus. Capillitium a delicate net-work of very slender threads, with no expansions at the angles; the lime mostly concentrated in one large, snow-white nodule at the center, a few very small, roundish nodules scattered through the net-work. Spores globose, very minutely warted, violaceous, 7–9 mic. in diameter.

Growing on old wood, mosses, etc.; a common species. Sporangium .4–.5 mm. in diameter, the stipe two or three times this length. *Tilmadoche compacta* Wingate. It is doubtful if *Tilmadoche columbina* Rost. belongs to this species. According to Lister, *Lepidoderma stellatum* Massee, is the same as this species, and if it be objected to the name that there is already a *Physarum compactum* Ehrenberg, it may have to be called *Physarum stellatum*.

b. Sporangium more or less irregular.

6. Physarum leucophæum Fr. Sporangium globose or depressed-globose, more or less irregular, the base never umbilicate, stipitate or subsessile; the wall a thin violaceous membrane, rugulose and iridescent, with a thin coat of small white scales and granules of lime, or sometimes nearly naked. Stipe variable in length, sometimes very short or quite obsolete, occasionally a few of them confluent, wrinkled, and sulcate, brown below, paler or whitish above. Capillitium a dense irregular net-work of slender tubules, more or less expanded at the angles; the nodules of lime white, small, roundish, or angular, few and scattered. Spores globose, very minutely warted, violaceous, 8-10 mic. in diameter.

Growing on old wood, bark, leaves, etc. The sporangium .5-.7 mm. in diameter, the stipe about the same length, or shorter, and sometimes wanting. The lime on the wall and in the capillitium is never abundant and sometimes extremely [Pg 92] scanty. Rostafinski's presentation of this species applies well to our specimens.

7. Physarum connexum Link. Sporangia subglobose, depressed, more or less irregular, sometimes confluent, stipitate, or subsessile; the wall a thin violaceous, or brownish membrane, rugulose, thickly covered with small white roundish scales of lime, which sometimes accumulate so as to make the surface rough and uneven. Stipe short, thick, rugulose, from snow white to smoky or sooty, especially toward the base, sometimes with a scanty calcareous hypothallus. Capillitium a loose net-work of tubules, much expanded at the angles; the nodules of lime small, white, rather numerous, ellipsoidal or fusiform, sometimes confluent and elongated. Spores irregularly globose, minutely warted, dark violaceous, 9-11 mic. in diameter.

Growing on old wood and bark. Sporangium .6-1.0 mm. in diameter, the stipe usually shorter than the diameter, sometimes very short; the lime-nodules about the thickness of the spores. This is a larger and rougher species than *P. leucophæum*, the sporangium is more often irregular and the spores darker colored. *P. confluens* and *P. connexum* of Link. See Plate XIV. Fig. 59.

8. Physarum compressum A. & S. Sporangium laterally compressed and much flattened, subreniform, stipitate or subsessile; the wall a thin violaceous or brownish membrane, rugulose, thickly covered with small white roundish nodules of lime, similar to those in the capillitium. Stipe short, brown or blackish at least below, sometimes pallid or grayish above, longitudinally rugulose. Capillitium of slender tubules, forming a loose net-work; the nodules of lime small, white, very numerous, roundish or ellipsoidal, often confluent end to end. Spores irregularly globose or angular, minutely warted, dark violaceous, 11–14 mic. in diameter.

Growing on old stalks and leaves of *Zea mays*. Sporangium variable, .6–1.0 mm. in breadth, the stipe 1 mm. or less in length; the lime nodules about the thickness of the spores. According to Saccardo this species is the same as *Physarum nephroedium* Rost.

9. Physarum polycephalum Schw. Sporangia confluent into a subspheric gyrose-complicate head, composed of [Pg 93] several to many laterally compressed, irregular, simple sporangia; the wall a thin, pellucid membrane, covered by a thin layer of minute scales of lime, white to yellow or greenish-yellow Stripes thin, flat, weak, and often prostrate, pale yellow, more or less connate, arising from a thin hypothallus. Capillitium of slender tubules forming a loose, irregular network, more or less expanded at the angles: the lime-nodules white or yellow, small, fusiform or by confluence elongated and sometimes branched. Spores globose, very minutely warted, violaceous, 8–10 mic. in diameter.

Growing on old bark, wood, leaves, etc. The sporangia rarely simple, usually confluent into a head of from four or five to fifteen or twenty, and sometimes more, simple sporangia; the stipes variable in length, long or short, rarely wanting. The gray form is *Didymium polymorphum* Mont., the yellow-green form *D. gyrocephalum* Mont. Sprengel considered this species the same as *Physarum compactum* Ehr., and it appears under this name in Schweinitz's *North American Fungi*; but Fries, who had seen specimens of both, disposed of them differently. See Plate XIV. Fig. 60.

10. Physarum didermoides Pers. Sporangia obovoid-oblong, stipitate, growing close together on a white membranaceous common hypothallus; the wall with a thick, white, outer layer of lime, easily

crumbling and falling away, leaving the sporangium dark gray; the inner membrane rather thick and firm, violaceous, with a closely adherent layer of granules of lime. Stipes very short, white, thin, and weak, each formed by a bit of membrane arising from the hypothallus. Capillitium a loose net-work of slender threads, bearing numerous roundish or irregular white nodules of lime. Spores irregularly or angularly globose, minutely warted, dark violaceous, 12–15 mic. in diameter.

Growing on wood, leaves, grass, etc. Sporangia .6–1.2 mm. in length by .4–.6 mm. in thickness, the stipe shorter than the sporangia. *Spumaria licheniformis* Schw., belongs here. This is a truly abnormal species of *Physarum*, so much so that Fries, in the *Summa Veg. Scand.* placed it by itself in a separate genus, *Claustria*. [Pg 94]

B. Sporangia sessile.

11. Physarum confluens Pers. Plasmodiocarp roundish, oblong or elongated, and by confluence branched and reticulate; the wall a thin, violaceous membrane, rugulose, with a thin, closely adherent layer of minute granules of lime, over which are scattered small, white, roundish nodules, which sometimes accumulate into a thick, pulverulent coat. Capillitium a loose net-work of tubules, widely expanded at the angles; the nodules of lime small, white, very numerous, roundish or ellipsoidal, by confluence elongated and irregular. Spores irregularly globose, minutely warted, dark violaceous, 9–11 mic. in diameter.

Growing on old wood, bark, leaves, etc. Plasmodiocarp .4–.5 mm. in thickness, varying from roundish to much elongated, creeping and reticulate. The sporangium before dehiscence is gray, whence Link's name, *Physarum griseum*; the loose pulverulent coating of lime easily falls away, leaving the sporangium dark colored, whence Rostafinski's name, *Physarum lividum*. The amount of lime on the wall and in the capillitium is variable.

12. Physarum luteolum Peck. Sporangia small, subglobose, sessile, closely gregarious; the wall a thin membrane, covered by a layer of small scales of lime, yellowish, inclining to tawny, in color, rupturing irregularly. Capillitium of slender tubules, forming a dense net-work of small meshes, scarcely expanded at the angles; the nodules of lime small, numerous, yellowish, roundish, or ellip-

soidal. Spores globose, nearly smooth, violaceous, about 10 mic. in diameter.

Growing on living leaves of *Cornus canadensis*, Adirondack Mountains, New York. I have not seen a specimen of this *Physarum*, but from Professor Peck's description and figure it seems to be a unique species.

13. Physarum thejoteum Fr. Sporangia very small, sessile, on a thin membranaceous hypothallus, closely crowded together and more or less connate, subobovoid or oblong, irregular from mutual pressure; the wall a thin violaceous membrane, closely covered with a thin layer of small irregular scales of lime, tawny or yellowish tawny in color, breaking up irregularly about the apex. Capillitium a loose irregu [Pg 95] lar net-work of slender threads, more or less expanded at the angles; the lime nodules small, tawny or yellowish, not numerous, ellipsoidal or fusiform, by confluence elongated and irregular. Spores globose, even, violaceous, 6–7 mic. in diameter.

Growing on old wood, mosses, etc. Sporangia .2–.4 mm. in diameter at the apex, densely packed and their walls grown together, approaching the aethalioid structure; the lime-nodules from one to two or three times the diameter of the spores in thickness. I have described my specimens, which are abundant, very carefully, and judge them to be referable to this species; if so, they show that the species should be kept apart from *Physarum virescens*. *Didymium nectriæforme* B. & C., is evidently this same species.

14. Physarum lateritium B. & R. Sporangia sessile, irregularly globose and gregarious, or by confluence more or less elongated and plasmodiocarp; the wall a thin violaceous membrane, rugulose and iridescent, closely covered with small irregular scales of lime, from testaceous or brick-red to bright red in color. Capillitium a dense irregular net-work of tubules, much expanded at the angles; the nodules of lime small, very numerous, roundish or angular, whitish or yellowish, sometimes tinged with red granules. Spores globose, very minutely warted, violaceous, 8–10 mic. in diameter.

Growing on old wood, sticks, leaves, etc. Sporangia .4–.6 mm. in diameter, by confluence sometimes much elongated; the lime-nodules two or three times the diameter of the spores in thickness.

Didymium lateritium B. & R. *Physarum inequale* Peck, is the same species. See Plate XIV. Fig. 61.

§2. Saxella. Lime in the capillitium abundant, the nodules large, angular or irregular, with pointed lobes and branchlets.

A. Sporangia stipitate.

15. Physarum imitans Racib. Sporangium depressed-globose, the base flattened or umbilicate, stipitate, erect or cernuous; the wall a thin violaceous membrane, with a [Pg 96] closely adherent layer of minute granules, over which are scattered rather large, roundish or irregular white scales of lime, splitting from the apex downward into a few irregular segments. Stipe short, thick at the base and tapering upward, longitudinally rugulose, from gray to brown or blackish, especially below. Capillitium a loose irregular network of tubules, widely expanded at the angles; the nodules of lime white, numerous, large, irregular, with pointed angles and lobes. Spores globose, very minutely warted, violaceous, 8–9 mic. in diameter.

Growing on old wood, mosses, etc. Sporangium .4–.5 mm. in diameter, the stipe about the same length or a little longer. The species superficially resembles the gray form of *Physarum nutans*, and quite likely is constantly overlooked on this account. Although I am not able to verify my reference, yet my specimens answer so well to the description of Raciborski that I am unwilling to invent a new name. See Plate XIV. Fig. 62.

16. Physarum ornatum Peck. Sporangium globose or depressed-globose, stipitate; the wall a thin yellowish membrane, covered with minute granules and small irregular scales of lime, yellow to orange in color. Stipe short, erect, blackish-brown, black at the base, longitudinally plicate, rising from a small hypothallus. Capillitium of tubules forming a rather dense net-work, with wide expansions at the angles; the nodules of lime large, numerous, yellow, irregular, sometimes confluently branched and reticulate. Spores globose, minutely warted, dark violaceous, 10–12 mic. in diameter.

Growing on old wood, bark, mosses, etc. Sporangium about .5 mm. in diameter, the stipe about the same length or shorter. *Physarum oblatum* McBride, can not be distinguished from this. Specimens of this species in the herbarium of Schweinitz are labeled *Physarum sulphureum*; this is without doubt a mistake.

17. Physarum gravidum Morgan, n. sp. Sporangium depressed-globose, the base umbilicate, stipitate; the wall a thin, violaceous membrane, brownish at the base, with a thin coat of small, white scales and minute granules of lime. Stipe long, erect, brown or reddish-brown, darker below, tapering [Pg 97] upward, expanding at the base into a small hypothallus. Capillitium of slender tubules forming a loose net-work, more or less expanded at the angles and for the most part filled with lime; the nodules white, slender, much elongated and branched, with pointed lobes and branchlets. Spores globose, very minutely warted, dark violaceous, 11–13 mic. in diameter.

Growing on old stalks of *Zea mays*. Sporangium .5–.6 mm. in diameter, the stipe about twice this length. The lower part of the capillitium is sometimes entirely filled with lime, so that the species approaches Badhamia in the structure of its capillitium.

18. Physarum leucopus Link. Sporangium globose, the base slightly flattened, stipitate; the wall a thin, violaceous membrane, with a white, pulverulent outer coat of minute granules of lime. Stipe short, thick, erect, snow-white, longitudinally rugulose, tapering upward, expanding at the base into small, white hypothallus. Capillitium a loose net-work of tubules, with wide expansions at the angles; the nodules of lime large, white, numerous, irregularly lobed and branched. Spores globose, very minutely warted, violaceous, 8–10 mic. in diameter.

Growing on old wood, leaves, etc. Sporangium .3–.4 mm. in diameter, the stipe about the same length as the diameter. Our specimens are a smaller form than the European, with smaller and smoother spores. Superficially the species resembles *Didymium squamulosum*, and it is *Didymium leucopus* of Fries, S. M.

19. Physarum glaucum Phillips. Sporangium globose, or the base slightly depressed, stipitate; the wall a thin, violaceous membrane, covered with minute, white granules and small roundish or irregular scales of lime. Stipe short, stout, erect, black, longitudinally wrinkled, expanding at the base into a small hypothallus. Capillitium of much-flattened tubules, forming a loose net-work, widely

expanded at the angles; the nodules of lime numerous, large, white, irregular, with pointed angles and lobes. Spores globose, very minutely warted, dark violaceous, 12–14 mic. in diameter.

Growing on old leaves: California. Sporangium .5-.7 mm. [Pg 98] in diameter, the stipe not longer than the diameter. This is quite a robust species, both externally and in the broad, flat tubules of the capillitium. See Plate XV. Fig. 64.

20. Physarum relatum Morgan, n. sp. Sporangium globose, the base umbilicate, stipitate, often cernuous; the wall a thin, violaceous membrane, rugulose and iridescent, covered with small, roundish or irregular white scales of lime. Stipe long, erect or inclined, rising from a thin hypothallus, tapering upward, white or cream color to ochraceous. Capillitium a dense net-work of tubules, more or less expanded at the angles, and almost entirely filled with white granules of lime, leaving only here and there short, slender empty spaces. Spores globose, nearly smooth, violaceous, 8–9 mic. in diameter.

Growing on old wood. Sporangium .5-.6 mm. in diameter, the stipe about twice this length. The capillitium is rigid, with the abundance of lime almost as in the genus *Badhamia*. Superficially the species much resembles *Cytidium globuliferum* or *Physarum compactum*, but the disposition of the lime on the wall and in the capillitium is altogether different. See Plate XIV. Fig. 63.

21. Physarum auriscalpium Cke. Sporangia subglobose, depressed, substipitate; the wall a hyaline membrane with a thin, closely adherent layer of minute granules of lime, over which are scattered large, irregular, orange-red scales of lime. Stipe very short, sometimes almost obsolete. Capillitium of tubules forming a loose net-work, with widely expanded angles, and mostly filled with orange granules of lime, only here and there short, slender, empty spaces. Spores globose, minutely warted, dark violaceous, 11–13 mic. in diameter.

Growing on rotten wood; South Carolina, Ravenel. Sporangia .6-.8 mm. in diameter, the stipe very short. Described in *Annals of the Lyceum of Natural History of New York*, June, 1877. So fine a species ought to be found again. Cooke's specimen was examined by Lister, *Mycetozoa*, p. 61.

EXPLANATION OF PLATE XIV.

- Fig. 56.—Craterium minimum B. & C. *a.* Sporangia × 5. *b.* Sporangium with lid × 90. *c.* Capillitium and spores × 500.
- Fig. 57.—Craterium maydis Morgan. *a.* Sporangia × 5. *b.* Sporangium × 90. *c.* Capillitium and spores × 500.
- Fig. 58.—Physarum obrusseum B. & C. *a.* Sporangia × 5. *b.* Sporangium × 90. *c.* Capillitium and spores × 500.
- Fig. 59.—Physarum connexum Link. *a.* Sporangia × 5. *b.* Sporangium × 90. *c.* Capillitium and spores × 500.
- Fig. 60.—Physarum polycephalum Schw. *a.* Sporangia × 5. *b.* Sporangia × 90. *c.* Capillitium and spores × 500.
- Fig. 61.—Physarum lateritium B. & C. *a.* Sporangia × 5. *b.* Sporangia × 90. *c.* Capillitium and spores × 500.
- Fig. 62.—Physarum imitans Racib. *a.* Sporangia × 5. *b.* Sporangium × 90. *c.* Capillitium and spores × 500.
- Fig. 63.—Physarum relatum Morgan. *a.* Sporangia × 5. *b.* Sporangia × 90. One divested of the wall and showing the rigid capillitium. *c.* Capillitium and spores × 500.

The Journal of the Cin. Soc. Natural History
Vol. XIX. Plate XIV.
Morgan on Myxomycetes.

B. Sporangia sessile.

22. Physarum plumbeum Fr. Sporangia small, globose or obovoid, sessile, on a narrow base, gregarious, sometimes close but

seldom confluent; the wall a thin violaceous mem [Pg 99] brane, with a very thin layer of small white scales and minute granules of lime, sometimes naked. Capillitium a loose net-work of slender tubules, with slight expansions at the angles; the nodules of lime white, numerous, more or less elongated, irregularly lobed and branched. Spores globose, even, violaceous, 7–9 mic. in diameter.

Growing on old leaves, sticks, etc. Sporangia .3–.4 mm. in diameter, quite regular in shape, attached by a narrow base, sometimes by a mere point, rarely confluent. The lime on the wall of the sporangium is rather scanty, sometimes altogether absent, and the nodules of lime in the capillitium are rather small. The species is figured by Micheli N. P. G. Tab. 96, Fig. 9. It is named by Fries S. M., III, p. 142. It is figured again by De Bary, *Die Mycetozoen*, Tafel I.

23. Physarum atrum Schw. Sporangia sessile, subglobose or oblong, by confluence, more or less elongated, bent or flexuous and branched; the wall a thin violaceous membrane, rugulose, covered by a wrinkled and reticulate layer of white granules of lime, which sometimes become thin or disappear. Capillitium a loose net-work of tubules, more or less expanded at the angles; the nodules of lime white, numerous, large, irregularly lobed and branched. Spores globose, very minutely warted, violaceous, 8–10 mic. in diameter.

Growing on old leaves, bark, grasses, etc.; apparently the most common of these three cinereous species. Sporangia .3–.5 mm. in thickness, some of them roundish or oblong, others elongated to several millimeters. The sporangium is often elegantly reticulate as observed by Schweinitz even when the lime is quite scanty. In Saccardo's *Sylloge* Berlese changed the name to *Physarum reticulatum*, but this is unnecessary, as the *Physarum atrum* of Fries is not a Myxomyces.

24. Physarum cinereum Batsch. Sporangia large, subglobose, sessile, gregarious, sometimes close and confluent; the wall a thin violaceous membrane, with a closely adherent layer of minute granules, over which are scattered irregular white scales of lime. Capillitium of tubules forming a loose net-work, with wide expansions at the angles; the nodules of lime numerous, white, very large, with pointed angles and lobes, by confluence often branched and reticulate, [Pg 100] and occasionally forming a pseudo-columella in the

center of the sporangium. Spores globose, minutely warted, dark violaceous, 10–13 mic. in diameter.

Growing on old wood, leaves, etc. The sporangia .4-.6 mm. in diameter, more or less irregular. The great abundance of lime in the capillitium and the large distinctly warted spores distinguish this species. *Physarum cinereum* of Persoon's Synopsis, *Didymium cinereum* of Fries' *Systema*. The only American specimens I have of this species are from Iowa (*McBride*) and from Nebraska (*Webber*).

25. Physarum virescens Ditm. Sporangia large, subglobose, irregular and unequal, sessile, gregarious, sometimes crowded, but not often confluent; the wall a thin membrane, violaceous, or in places yellowish, with a dense layer of yellow or greenish-yellow scales and granules of lime. Capillitium a loose net-work of tubules, with wide expansions at the angles; the nodules of lime large, numerous, yellow or greenish-yellow, more or less elongated, lobed, and branched. Spores globose or somewhat irregular, very minutely warted, violaceous, 9–11 mic. in diameter.

Growing on old leaves, mosses, etc. Sporangia .5-.8 mm. in diameter, occasionally by confluence more elongated. Though found in all parts of the country, the species seems rare. This is not the *Physarum virescens* described by Rostafinski.

26. Physarum rubiginosum Fr. Sporangia subglobose, sessile, gregarious; the wall a thin hyaline membrane, thickly covered with large irregular scales of lime, orange to red or dark red in color, breaking up irregularly. Capillitium of hyaline tubules, forming a loose irregular net-work, more or less expanded at the angles; the nodules of lime large, angular, and irregular, sometimes confluent, orange to dark red in color. Spores globose, very minutely warted, dark violaceous, 9–11 mic. in diameter.

Growing on old wood, leaves, mosses, etc. Sporangia .6-.8 mm. in diameter. *Physarum fulvum* Fries S. M., III, p. 143. A rare species. It should not be confounded with *Physarum lateritium*. [Pg 101]

27. Physarum serpula Morgan, *n. nom.* Plasmodiocarp roundish or oblong to much elongated, bent, annular and flexuous, sometimes by confluence branched and reticulate; the wall a firm yellowish membrane, with a thin, rough, closely adherent coat of granules

of lime, dull ochre to lemon-yellow and orange in color. Capillitium a dense net-work of tubules, for the most part filled with lime, only here and there short, slender, empty spaces; the nodules large, numerous, white or yellow, angular and with pointed lobes and branchlets. Spores globose, minutely warted, dark violaceous, 9–11 mic. in diameter. See Plate XV. Fig. 65.

Growing on leaves, bark, lichens, etc. Plasmodiocarp .3-.4 mm. in thickness and of varying length. This species is in the herbarium of Schweinitz, at Philadelphia, with the name *Physarum reticulatum*; it is described by George Massee as *Physarum gyrosum*; by Lister it is incorporated with several other species under *Badhamia decipiens*.

28. Physarum contextum Pers. Sporangia sessile and closely crowded together, roundish or more or less elongated, flexuous and complicate, the apex plane or impressed; the wall a firm yellowish membrane, covered by a thick pulveraceous layer of lime, white, ochraceous or yellow, easily crumbling and breaking up. Capillitium a loose net-work of tubules, much expanded at the angles; the nodules of lime very large, white or yellow, numerous, angular, and irregular, by confluence lobed and branched, sometimes massed together in the center of the sporangium. Spores globose, minutely warted, dark violaceous, 10–13 mic. in diameter.

Growing on bark, leaves, mosses, etc. Sporangia with a width of .3-.5 mm. and varying in length from .5 mm. to 1 or 2 mm. The sporangia are often so much crowded as to appear to be grown together. *Diderma ochroleucum* B. & C. belongs to this species. *Physarum conglomeratum* Fr. is a closely related species, with smaller and smoother spores. I have not met with this.

29. Physarum diderma Rost. Sporangia large, irregularly globose or oblong, sessile, but without a hypothallus, closely crowded together and sometimes confluent. The wall composed of two distinct and separate layers; the outer a [Pg 102] thick, uneven, crustaceous, snow-white layer of lime; the inner a thin, violaceous membrane, cinereous from the adherent granules of lime, or free from them, and iridescent. Capillitium of tubules forming a loose net-work, with wide expansions at the angles; the nodules of lime numerous, snow-white, large, irregular, with pointed angles and lobes, some-

times confluent in the center of the sporangium. Spores globose, minutely warted, dark violaceous, 9–10 mic. in diameter.

Growing on wood, bark, and mosses. Sporangia .8–1.0 mm. in diameter, more or less irregular. The wall of the sporangium is exactly like that of certain species of *Diderma*. This species must be rare, as I have met with it but twice in ten years, and I am not aware that it has ever been found by any one else.

VIII. FULIGO Haller. Aethalium a compound plasmodiocarp; the component sporangia branching and anastomosing in every direction, complicate and grown together; the walls of the sporangia a thin membrane, coated with minute, roundish granules of lime. Capillitium of tubules forming a net-work of irregular meshes, more or less expanded at the angles, the tubules containing in greater or less abundance irregular nodules of lime. Spores globose or sometimes ellipsoidal, violaceous.

The genus is readily distinguished from *Spumaria* by the round granules of lime upon the walls of the sporangia.

§1. Aethalium Link. Aethalia large; the lime in the capillitium scanty, the nodules small, ellipsoidal, or fusiform.

a. Aethalium with a thick fragile common cortex.

1. Fuligo rufa Pers. Plasmodium a large soft mass with a peculiar odor and golden yellow in color. Aethalium very large, pulvinate, orbicular, elongated, or quite irregular, extremely friable, the surface tawny or ferruginous to ochraceous and whitish. The long narrow, sinuous sporangia closely compacted, entirely grown together and inseparable, covered by a thick common cortex, and seated on a much [Pg 103] thickened hypothallus; walls of the sporangia a thin pellucid membrane, coated by a thin layer of white granules of lime. Capillitium of very slender tubules, extending across from wall to wall, sparingly branched and scarcely forming a network, not at all or only slightly expanded at the angles; the tubules for the most part empty, here and there with slight fusiform or elongated swellings containing granules of lime, occasionally bearing roundish or ellipsoidal nodules of larger size. Spores globose, nearly smooth, violaceous, 6–9 mic. in diameter.

Growing on old trunks in woods in great abundance from early Spring to Winter. Aethalium 3–6 or sometimes many centimeters in extent and 1–2 cm. in thickness. The common cortex and the hypothallus are a millimeter or more in thickness; they are composed of successive layers of thin plates of membrane coated with granules of lime.

b. Aethalium naked, i. e., without a common cortex.

2. Fuligo violacea Pers. Plasmodium a soft effused mass, dark red or wine-colored. Aethalium large, pulvinate or effused, orbicular or more or less elongated and irregular, the surface minutely pitted and perforate, furnished with a scanty layer of lime, whitish or yellowish to brick-red in color, leaving naked purple and violet spots and patches, seated on a thin membranaceous brick-red hypothallus. Sporangia long, narrow, and sinuous, closely packed together; the walls a thin violaceous membrane, rugulose and iridescent, with scattered granules, or nearly destitute of lime. Capillitium of slender violet tubules, forming a loose net-work, with slight expansions at the angles; the tubules with numerous rather large vesicular expansions, ellipsoid or fusiform in shape, and scantily furnished with lime. Spores globose, nearly smooth, pale vinous, 6–8 mic. in diameter.

Growing on old trunks in woods; not uncommon in this region. Aethalium 1–3 or more centimeters in extent, and 5–10 mm. in thickness. The vesicles of the capillitium vary from 15–30 or sometimes to 50 mic. in diameter, their inner surface is usually coated by a single layer of granules of lime, they are rarely filled with lime and sometimes are naked entirely; when dry many of them are to be found collapsed. See Plate XV. Fig. 66. [Pg 104]

3. Fuligo flava Pers. Plasmodium effused lemon-yellow. Aethalium mostly effused, irregular, the surface reticulate, pitted and perforate, entirely naked, pale yellow to lemon-yellow and greenish-yellow, the hypothallus thin or scarcely evident. Sporangia laterally much compressed, flexuous, and gyrose, not everywhere grown together, but forming a dense reticulum; the walls a thin, pellucid membrane, with a dense layer of lemon-yellow granules of lime. Capillitium of short and very slender tubules, sparingly branched and scarcely forming a net-work, not expanded at the angles; the

tubules very scantily furnished with lime, in scattered, small, fusiform nodules, white or lemon-yellow. Spores globose, very minutely warted, violaceous, 7–9 mic. in diameter.

Growing on mosses, old leaves, sticks, etc.; not common. Aethalia in irregular patches 2–4 cm. or more in extent, sometimes almost reduced to a simple plasmodiocarp. This species furnishes a clear notion of the structure of the aethalium in the other species, on account of the sporangia being but loosely compacted and not entirely grown together. The *Fuligo vaporaria* Pers., of the green-houses and gardens I have never seen; the *Mucor septicus* Linn., was thought to be the plasmodium of this. Linnæus's description is simply "*Mucor unctuosus flavus.*" See Plate XV. Fig. 67.

§2. Aethaliopsis Zopf. Aethalium small; lime abundant in the capillitium, the nodules numerous and large, angular and irregular.

4. Fuligo muscorum A. & S. Plasmodium effused, golden yellow. Aethalium small, subpulvinate, irregular, the surface furnished with scattered, irregular scales of lime, whitish or ochraceous to golden yellow in color, arising from a thin, white, membranaceous hypothallus. Sporangia closely packed and grown together; the walls a thin, violaceous membrane, rugulose, with a thin, closely adherent layer of granules of lime. Capillitium a loose net-work of tubules, widely expanded at the angles; the tubules for the most part filled with lime, the nodules white or yellowish, numerous, very large, angular and irregular, sometimes confluent with pointed lobes and branchlets. Spores irregularly globose, minutely warted, dark violaceous, 9–11 mic. in diameter. [Pg 105]

Growing on leaves, twigs, mosses, etc. Aethalium from 2 or 3 mm. to a centimeter or more in extent. I have a specimen of *Fuligo simulans* Karsten, from Karsten himself; it is identical with my specimens of *Fuligo ochracea* Peck. There could be no better representation of these specimens made at that time than the description and figure of *Fuligo muscorum* A. & S., in the *Conspectus*.

5. Fuligo cinerea Schw. Plasmodium milk-white, changing to cinereous. Aethalium effused, variable in extent, the surface rugulose and perforate, white, the hypothallus thin or scarcely evident. Sporangia variously contracted and grown together, forming a dense reticulum; the walls a thin pellucid membrane, with a thick white

outer layer of granules of lime. Capillitium a loose net-work of tubules, widely expanded at the angles, the tubules for the most part filled, with lime, the nodules white, numerous, very large, angular, and irregular, lobed and branched. Spores globose or oval, minutely warted, dark violaceous, 10–15 × 10–12 mic.

Growing on old leaves, herbaceous stems, etc. I find it most abundantly about the horse barn, upon the old straw and manure, sometimes running out onto the green herbage. Aethalium from a few millimeters to several centimeters in extent. Upon the testimony of Dr. Geo. A. Rex this is both *Enteridium cinereum* and *Lachnobolus cinereus* of Schweinitz's *North American Fungi* as represented in his herbarium. It is *Physarum ellipsosporum* of Rostafinski. It is no doubt also *Aethaliopsis stercoriformis* Zopf. See Plate XV. Fig. 68.

IX. BADHAMIA Berk. Sporangia large, subglobose or obovoid, sometimes depressed, substipitate or sessile; the wall a thin membrane, with an outer layer of minute roundish granules of lime, irregularly dehiscent. Stipe poorly developed, sometimes a mere strip of the hypothallus, often wanting. Capillitium of thick tubules, attached on all sides to the wall of the sporangium, combined into a net-work of large meshes, more or less expanded at the angles; the tubules containing minute roundish granules of lime throughout their whole extent. Spores large, subglobose, dark violaceous. [Pg 106]

The peculiar character of this genus is that the granules of lime are distributed along the whole interior of the tubules of the capillitium; this makes the net-work rigid, and on this account a firmer support for the wall of the sporangium.

1. Badhamia capsulifera Bull. Sporangia subglobose or obovoid, sessile, on a thin pallid or yellowish hypothallus, which sometimes sends out narrow bands or strings of membrane of variable length, bearing sporangia singly or in clusters. Wall of the sporangium a thin pellucid membrane, mostly even or somewhat rugulose and iridescent, coated by a very thin layer of white granules of lime. Capillitium of rather slender tubules, forming an open net-work of very large meshes, only slightly expanded at the angles; the tubules coated within by a very thin layer of white granules of lime. Spores subglobose or obovoid, adhering together in clusters of six to twen-

ty or more, distinctly warted on the outer exposed surface, dark violaceous, 10–13 mic. in diameter.

Growing on old bark, leaves, etc. Sporangia .8–1.4 mm. in diameter. *Badhamia hyalina* and *B. capsulifera* of Rostafinski's monograph are here included together; he distinguished the former by the "sporangia in clusters always exactly globose," a distinction first made by Chevallier; otherwise the characters are the same in both.

2. Badhamia utricularis Bull. Sporangia subglobose or obovoid, sessile, on a thin pallid or yellowish hypothallus, which often separates into narrow strips and strings of membrane of variable length, bearing the sporangia singly or in clusters. Wall of the sporangium a thin violaceous membrane, rugulose and iridescent, shining with purple, blue, and violet tints, with a thin layer of white granules of lime. Capillitium of thick tubules, forming an open net-work of large meshes, more or less expanded at the angles, the tubules coated within by a thin layer of granules of lime. Spores subglobose, minutely warted, dark violaceous, 10–13 mic. in diameter.

Growing on old wood, bark, herbaceous stems, etc. Sporangia .5–1.0 mm. in diameter, usually growing in clusters, sometimes suspended by the strings of membrane. Rostafinski's distinction between this and the preceding species in [Pg 107] regard to the spores holds good so far as my specimens are concerned. *Badhamia magna* Peck, I have never seen. George Massee includes all four of these species in one composite species, which he names *Badhamia varia*.

3. Badhamia papaveracea B. & Rav. Sporangia subglobose or obovoid, substipitate or sessile, growing close together; the wall a thin violaceous membrane, rugulose and iridescent, with a very thin coat of white granules of lime. Stipe very short, brown or blackish, sometimes reduced to merely a thickened blackish base to the sporangium. Capillitium of thick tubules, forming an open net-work of large meshes, more or less expanded at the angles; the tubules with an inner lining of very minute white granules of lime. Spores adhering together in clusters of six to twenty, each spore subobovoid, the free portion more distinctly warted, 10–12 mic. in diameter.

Growing on old wood, bark, etc. Sporangia .6–1.0 mm. in diameter. Readily distinguished by its black base or black stipe and the elegant clusters of its spores, which stick together most persistently. See Plate XV. Fig. 69.

4. Badhamia orbiculata Rex. Sporangia much depressed, orbicular or somewhat irregular, umbilicate often both above and below, gregarious, sometimes growing close together and confluent, stipitate or sessile. The wall a thin pellucid membrane, with a thin layer of minute granules of lime, which are sometimes raised into small scales and fine ridges. Stipe very short, black, sometimes reduced to merely a blackish base to the sporangium. Capillitium of thick tubules, forming a scanty irregular net-work, with wide expansions at the angles; the tubules filled with white granules of lime. Spores subglobose, very minutely warted, dark violaceous, 12–15 mic. in diameter.

Growing on old bark, herbaceous stems, etc. Sporangia .4-.8 mm. in diameter, sometimes by confluence larger. This species seems near *Badhamia verna* Smfdt, but the latter everywhere is described as sessile, while in the former the short black stipe is nearly always distinguishable.

5. Badhamia affinis Rost. Sporangium hemispherical, or much depressed, the base flattened or umbilicate, stipitate, [Pg 108] erect or often cernuous; the wall a thin pellucid membrane, coated with minute white granules of lime, which are frequently raised into scales and ridges. Stipe short, erect or bent at the apex, black, expanding at the base into a small hypothallus. Capillitium of thick tubules, forming an open net-work of large meshes, more or less expanded at the angles; the tubules filled with white granules of lime. Spores subglobose, minutely warted, dark violaceous, 14–18 mic. in diameter.

Growing on mosses and upon the bark of maple trunks. Sporangium .6–1.0 mm. in diameter, the stipe about the same length. Rostafinski's description is based upon a specimen found in Chili, South America, by Bertero; it is recorded in this country by Peck. I find it in some seasons quite abundant. The spores are very large, in some specimens averaging 17 mic. See Plate XV. Fig. 70.

6. Badhamia decipiens Curtis. Sporangia gregarious, sessile, globose, oval or oblong, by confluence sometimes more elongated; the wall a somewhat thickened and firm yellow or yellow-brown mem-

brane, covered with large, thick scales of lime, tawny to golden yellow or orange in color. Capillitium of thick tubules, forming an open network, more or less expanded at the angles; the tubules filled throughout with yellow granules of lime. Spores globose, very minutely warted, lilac, 10–12 mic. in diameter. See Plate XV. Fig. 71.

Growing on old wood and bark. Sporangia .6–1.0 mm. in length by .6–.7 mm. in thickness. My specimens were determined by Dr. George A. Rex by comparison with a specimen from Curtis in the herbarium of the Philadelphia Academy of Sciences. This species should not be confused with what we have described as *Physarum serpula*. *Physarum chrysotrichum* B. & C., is no doubt the same thing. *Badhamia nitens* Berk., which is also golden yellow, has not yet been found in this country; it will readily be distinguished from the present species by its clustered spores.

7. Badhamia panicea Fr. Sporangia sessile, subglobose or oblong, more or less irregular, gregarious; the wall a thin, pellucid membrane, covered with large, irregular, very thick, white scales of lime. Capillitium of thick tubules, forming a [Pg 109] loose net-work of rather small meshes, with wide expansions at the angles; the tubules filled with white granules of lime, sometimes confluent toward the base of the sporangium. Spores subglobose, very minutely warted, dark violaceous, 11–14 mic. in diameter.

Growing on old wood, bark, leaves, etc. Sporangia .8–1.6 mm. in length, with a thickness of .7–1.0 mm. This species appears to be rare; the only specimens known to me in this country I have from Professor Thos. A. Williams, of South Dakota; they are identical with European specimens received from Lister. *Physarum paniceum* Fries, S. M., III, p. 141; it approaches *Physarum cinereum* Batsch.

8. Badhamia lilacina Fr. Sporangia globose or obovoid, sessile or rarely substipitate, closely crowded together on a thin, brownish hypothallus; the wall a firm, hyaline membrane, with a thick, smooth, continuous outer-layer of lime, varying in color from gray-white or drab to lilac and flesh color. Capillitium of very thick tubules, forming a dense net-work of small meshes; the tubules stuffed with granules of lime, which are white or colored somewhat as those in the wall, often confluent in the center of the sporangium.

Spores globose, minutely warted, dark violaceous, 11–14 mic. in diameter.

Growing on wood, leaves, mosses, etc. Sporangium .5-.7 mm. in diameter. The outer crustaceous layer of lime on the wall crumbles and falls away, as in some species of *Diderma*. The white form is *Diderma concinnum* B. & C.; the lilac or flesh-colored form is *Physarum lilacinum* of Fries, S. M., p. 141. I have seen it colored only white and drab. Under a high magnifying power the sculpturing of the spores is seen to be peculiar.

X. SCYPHIUM Rost. Sporangium obovoid to oblong-obovoid, stipitate or subsessile; the wall a thickened, brownish membrane, the surface entirely naked or only the upper portion covered with granules of lime, breaking up irregularly about the apex. Stipe variable in length, arising from a common hypothallus and prolonged within the sporangium as a columella. Capillitium of thick tubules, proceeding from numerous points of the columella and forming a dense net [Pg 110] work; the tubules filled with lime throughout their whole extent. Spores large, subglobose, dark reddish-brown.

This genus differs from *Badhamia* by the columella which gives origin to the capillitium. The sporangia in the species composing it, resemble those of *Craterium*, and to this genus they are referred by Massee, but the capillitium is that of *Badhamia*.

1. Scyphium rubiginosum Chev. Sporangia gregarious, obovoid, stipitate; the wall a thickened reddish-brown membrane, the upper part covered by a thin layer of white granules of lime, the lower basal portion naked, strongly venulose and more persistent. Stipe long, erect, reddish-brown, expanding at the base into a brown hypothallus, prolonged within the sporangium to more than half its height as a columella. Capillitium of thick tubules, forming a dense net-work of small meshes; the tubules stuffed with white granules of lime. Spores irregularly globose, minutely warted, dark reddish-brown, 12–15 mic. in diameter.

Growing on old wood, mosses, etc. Sporangia .6-.8 mm. in height by .5-.6 mm. in diameter, the stipe from once to twice the height of the sporangium. This is *Physarum rubiginosum* Chevallier, *Flor de Paris*. It is also *Craterium obovatum* Peck. See Plate XV. Fig. 72.

2. Scyphium curtisii Berk. Sporangia oblong-obovoid, stipitate or subsessile, usually growing in clusters; the wall a thick, firm, reddish-brown membrane, venulose and reticulate, nearly destitute of lime. Stipes variable, commonly very short, sometimes confluent, arising from a brown hypothallus, prolonged within the sporangium to about half its height. Capillitium of thick tubules, forming a dense network of small meshes; the tubules stuffed with white granules of lime. Spores irregularly globose, minutely warted, dark reddish-brown, 12–15 mic. in diameter.

Growing on old wood, leaves, grass, etc. Sporangium .4-.7 mm. in height by .3-.4 mm. in diameter, the stipe often reduced to a mere point or cushion on the hypothallus, and varying thence to nearly the length of the sporangium. The sporangium is narrower than in the preceding species, and the brown wall is usually without granules of lime. It is *Didymium curtisii* Berk. Rostafinski and Massee both preserve it distinct from *S. rubiginosum.* See Plate XV. Fig. 73.

EXPLANATION OF PLATE XV.

- Fig. 64.—Physarum glaucum Phillips, *a.* Sporangia × 5. *b.* Sporangium × 90. *c.* Capillitium and spores × 500.
- Fig. 65.—Physarum serpula Morgan, *a.* Plasmodiocarp × 5. *b.* Piece of plasmodiocarp × 90. *c.* Capillitium and spores × 500.
- Fig. 66.—Fuligo violacea Pers. *a.* Aethalium natural size. *b.* Capillitium and spores × 500.
- Fig 67.—Fuligo flava Pers. *a.* Portion of an aethalium × 5. *b.* Capillitium and spores × 500.
- Fig. 68.—Fuligo cinerea Schw. *a.* Portion of aethalium × 5. *b.* Capillitium and spores × 500.
- Fig. 69.—Badhamia papaveracea B. & Rav. *a.* Sporangia × 5. *b.* Sporangium together with transverse section × 90. *c.* Capillitium and spores × 90. *d.* Portion of capillitium with clustered spores × 500.
- Fig. 70.—Badhamia affinis Rost. *a.* Sporangia × 5. *b.* Sporangia × 90, one with section showing capillitium. *c.* Capillitium and spores × 500.

- Fig. 71.—Badhamia decipiens Curtis, *a.* Sporangia × 5. *b.* Sporangia × 90. *c.* Section of sporangium showing capillitium. *d.* Capillitium and spores × 500.
- Fig. 72.—Scyphium rubiginosum Chev. *a.* Sporangia × 90. *b.* Sporangia × 90, with section showing capillitium. *c.* Capillitium and spores × 500.
- Fig. 73.—Scyphium curtisii Berk. Sporangia × 5.

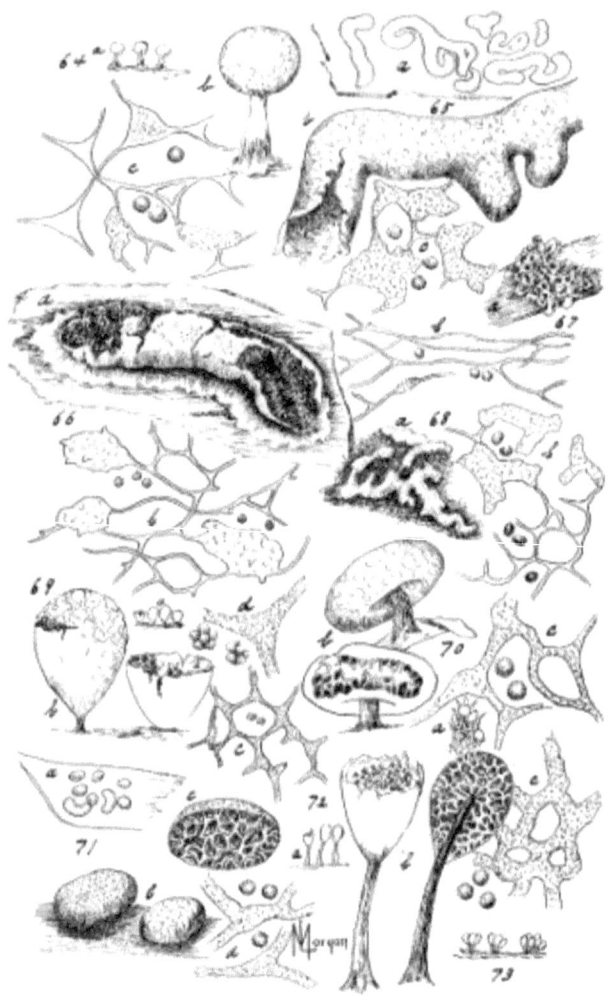

The Journal of the Cin. Soc. Natural History
Vol. XIX. Plate XV.
Morgan on Myxomycetes.

www.ingramcontent.com/pod-product-compliance
Lightning Source LLC
Chambersburg PA
CBHW031429210526
45464CB00005B/2113